GEOTECHNICS AND HERITAGE

GEOTECHNICS AND HERITAGE

Geotechnics and Heritage

Editors

Emilio Bilotta, Alessandro Flora, Stefania Lirer & Carlo Viggiani
University of Naples Federico II, Napoli, Italy

CRC Press
Taylor & Francis Group
Boca Raton London New York

CRC Press is an imprint of the
Taylor & Francis Group, an **informa** business
A BALKEMA BOOK

Cover photo
Description: Tree mirroring in water pond
Copyright: Trevi Group

CRC Press
Taylor & Francis Group
6000 Broken Sound Parkway NW, Suite 300
Boca Raton, FL 33487-2742

First issued in paperback 2019

© 2013 by Taylor & Francis Group, LLC
CRC Press is an imprint of Taylor & Francis Group, an Informa business

Typeset by V Publishing Solutions Pvt Ltd., Chennai, India

ISBN-13: 978-1-138-00054-4 (hbk)
ISBN-13: 978-0-367-37997-1 (pbk)

**Visit the Taylor & Francis Web site at
http://www.taylorandfrancis.com**

**and the CRC Press Web site at
http://www.crcpress.com**

MIX
Paper from
responsible sources
FSC FSC® C013985

Printed in the United Kingdom
by Henry Ling Limited

Geotechnics and Heritage – Bilotta, Flora, Lirer & Viggiani (eds)
© 2013 Taylor & Francis Group, London, ISBN 978-1-138-00054-4

Table of contents

Geotechnics and Heritage – Bilotta, Flora, Lirer & Viggiani (eds)
© 2013 Taylor & Francis Group, London, ISBN 978-1-138-00054-4

Preface

The Technical Committee on Preservation of Monuments and Historic Sites was established by the International Society of Soil Mechanics and Geotechnical Engineering in 1981 with the mark TC19, and renamed TC301 in 2010. The Committee is supported by the Italian Geotechnical Society (AGI); it has been chaired in the past by Jean Kerisel, Arrigo Croce, Ruggiero Jappelli.

Edmund Burke, in his "Reflection on the revolution in France", states as early as in 1790: "*People will not look forward to posterity, who never look backward to their ancestors*". And Lenin writes in the early XX century: "*Citizens, don't touch even a stone. Protect your monuments, the old mansions. They are your history, your pride*". Besides being so important, conservation is also one of the most challenging problems facing modern civilization. It involves a number of factors belonging to different fields (cultural, humanistic, social, technical, economical and administrative), intertwining in inextricable patterns. The complexity of the topic is such that it is difficult to imagine guidelines or recommendations summarizing what should be done and prescribing activities to carry on, intervention techniques, design approaches.

Instead of this ambitious undertaking, the Committee resolved to produce this volume collecting a number of relevant case histories concerning the role of Geotechnical Engineering in the preservation of monuments and historic sites, in addition to the Proceedings of the two International Symposia organized by the Committee in Napoli in 1994 and 2013. It is offered to the geotechnical engineers dealing with monuments and historic sites, as a collection of paradigmatic examples which may suggest an approach rather than a solution.

The lovely picture on the cover of this book illustrates at the best the concept of ground-monument system; it seems to suggest that a majestic tree must be based on a similarly majestic underground structure. It is hence evident that geotechnical engineers may play a significant role in conservation. We hope that this volume will contribute to such an undertaking.

The TC301 of ISSMGE

Geotechnics and Heritage – Bilotta, Flora, Lirer & Viggiani (eds)
© 2013 Taylor & Francis Group, London, ISBN 978-1-138-00054-4

Sponsor

This book has been printed with financial support from

Geotechnics and Heritage – Bilotta, Flora, Lirer & Viggiani (eds)
© 2013 Taylor & Francis Group, London, ISBN 978-1-138-00054-4

Cultural heritage and geotechnical engineering: An introduction

C. Viggiani
Emeritus Professor of Geotechnics, University of Napoli Federico II, Napoli, Italy
Chairman, TC301, ISSMGE

ABSTRACT: One of the fields of activity of Civil Engineering is the maintenance of existing constructions and infrastructures; among these, the Cultural Heritage. In particular, Geotechnical Engineering plays a significant role in a number of relevant cases.

The conservation of Heritage is one of the most challenging problems facing modern civilization. It involves a number of factors belonging to different fields (cultural, humanistic, social, technical, economical and administrative), intertwining in inextricable patterns.

From the point of view of an engineer, the peculiarity of this type of intervention is the requirement of respecting the integrity, besides guaranteeing the safe use. This requirement is analysed and discussed. It is concluded that the development of a shared culture between engineers and other professionals such as archaeologists, art historians and architects is a necessary condition for a successful conservation. To become sufficient too, the conservation culture should be spread widely and become a common sentiment amongst the majority of people.

1 INTRODUCTION

Geotechnical Engineering is part of Civil Engineering, the oldest branch of Engineering. As early as 30 centuries BC Egyptian engineers conceived and directed large constructions, as the great pyramids, the river Nile regulation and huge irrigation projects.

During the centuries, civil engineers have used the resources of Nature for the progress of mankind. At present, however, they are generally considered to operate in a "mature" sector, where mature is just an euphemism meaning outdated, behind the times, out of the mainstream of modernity. Till a few decades ago, the public perception of civil engineer was the man of progress; at present, this image is gradually changing into that of a cementifier, dangerous for the environment, operating in a routine sector of activity not worthy of investing financial and human resources; these on the contrary are reserved for fields such as aerospace, informatics, biomedicine.

We all know that this picture is utterly in error. According to the American Peoples Encyclopedia, Engineering "… applies scientific knowledge to the practical problems of creating, operating and maintaining structures, devices and services". Civil Engineers have thus still a fundamental role not only (i) in constructing new structures and (ii) in dealing with new problems, but also (iii) in taking care of the existing infrastructures and heritage.

As for new structures, just to quote a few examples, structures such as the Messina Strait suspension bridge, with its 3.3 km span (fig. 1), or skyscrapers such as the Burji Khalifa, with its 830 m height (fig. 2), will continue to require high level knowledge. Less visible (there is no glory in foundations, said Terzaghi), but probably equally demanding and more important and widespread are other works, as for instance the shallow urban tunnels for transportation and other purposes that are being excavated under all the great cities of the world.

New problems are continuously posed to civil engineers by the evolution of Society. For instance, environmental problems were not included in the engineering curricula when I was a student (admittedly, very long ago), in spite of them being described at the best as early as in Shakespeare's Hamlet. *"This goodly frame, the earth, seems to me a sterile promontory, this most excellent canopy, the air, look*

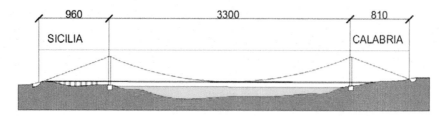

Figure 1. Sketch of the intended suspension bridge across Messina Strait, Italy.

Figure 2. The Burj Khalifa, the tallest building in the world.

you, this brave overhanging firmament, this majestical roof fretted with golden fire, why, it appears no other thing to me than a foul and pestilent congregation of vapours" says Hamlet entering the scene at the beginning of the second act.

We are used to associate sea gulls with blue skies and uncontaminated seas (fig. 3); but nowadays, on the contrary, they often deal with things such as solid urban wastes (fig. 4). Seagulls undertake a dirty job, just as we civil engineers do in addition to conceiving and constructing suspension bridges, skyscrapers, tunnels.

At a first glance, taking care of (maintaining, upgrading, reinforcing) existing structures, devices and services appears a minor field of activity; but there are many arguments proving the contrary. Apart from ordinary maintenance, puzzling problems are posed by new destinations of the existing structures or by more demanding safety requirements (for instance, in connection with the seismic safety of existing dams).

Figure 3. Blue skies and uncontaminated seas.

Figure 4. One of the present jobs of the seagulls.

Within this sector, there is the sub sector of taking care of (maintaining, preserving, conserving, restoring, improving) monuments and historic sites. Again, an apparently minor sub sector, and again, on the contrary, one of the utmost importance—may be the most significant contribution that Civil Engineering can give to mankind in our time.

Edmund Burke, in his famous "Reflection on the revolution in France", states in 1790: "*People will not look forward to posterity, who never look backward to their ancestors*". And Lenin writes: "*Citizens, don't touch even a stone. Protect your monuments, the old mansions. They are your history, your pride*". Both a reactionary and a revolutionary acknowledge the importance of the heritage and, oddly, the revolutionary seems to focus on the past and the reactionary on the future!

2 ITALIANS DO IT BETTER?

Italy has a long history of conservation; there is a widespread belief that the national heritage is very important. Let us examine a few quotations: "*According to UNESCO's estimates, Italy has between 60 and 70 per cent of the world's cultural assets*" (Eurispes Report, 2006). "*72% of Europe's cultural heritage is to be found in Italy and as much as 50% of the world's*" (Berlusconi, press conference held in London on 10 September 2008). According to a Sicilian minister, "*60 per cent of the world's cultural assets are in Italy and of these, 60 per cent are in Magna Grecia and of these last ones 60 per cent are in Sicily*". According to the councilor responsible for culture in the Regione Toscana, "*Italy alone has 60% of the world's cultural assets, but 50% of those Italian cultural assets are concentrated in Tuscany*"; according to the Deputy Mayor of Rome, Rome by itself "*has 30–40% of the world's cultural assets*". If we add all the percentages together, it would appear that Italy somehow manages to encompass more than 100 per cent of the planet's cultural heritage!

Obviously these "UNESCO statistics" do not exist, and the figures, invariably inconsistent with one another, are shamelessly concocted on various occasions; perhaps the symptom of national pride, but certainly of ill-conceived superficiality. Yet Italy does have a very significant position because of its cultural heritage, whose central importance, however, is not based on its quantity but rather on its quality. There are three different reasons for this: first, the time-honored harmony between the city and the wider landscape; second, the spread of this heritage throughout the country and down to the smallest towns and villages; third, the continuity in the use of churches, mansions, statues and paintings. Italian museums only contain a small portion of the country's artistic heritage, which is spread throughout the cities and the countryside: within this context—the product of many centuries of accumulated wealth and civilization—the whole is far more than the sum of its parts.

There is, however, a fourth factor at play, which is no less important: the "Italian model" of conservation of cultural heritage. Long before Italian unification, the Italian states formulated rules and set up public institutions to regulate and engage in this area of activity. Italy (as a state) was the first country to include the preservation of its landscapes and its cultural heritage amongst the founding principles of its Constitution. Article 9 of the Italian Constitution (which came into effect on January 1st, 1948) states: "The Republic promotes the development of culture and scientific and technical research. It protects the landscape and the Nation's historical and artistic heritage." (in the Italian original: *La Repubblica promuove lo sviluppo della cultura e la ricerca scientifica e tecnica. Tutela il paesaggio e il patrimonio storico e artistico della Nazione*). The Constituent Assembly arrived at this formulation after a long debate and eleven different proposed texts. Members of all parties contributed to the final wording, particularly the communist Concetto Marchesi, a professor of Latin from Sicily who had been rector at the University of Padua, and a very young Christian-democrat, later to be prime minister, Aldo Moro.

3 GEOTECHNICS AND HERITAGE

There are quite a number of monuments, monumental buildings, historical cities and sites affected by geotechnical risks of various types, so that Geotechnical Engineering is called to play a very important role in their safeguard. Just to quote some examples, the village of Terra Murata (Walled Land) in the island of Procida in the bay of Napoli, is a settlement dating back to the X century and includes the Abbey of S. Michele and a number of churches. The Abbey and other churches are exposed to the risk of collapse by the action of the sea, undermining the steep cliffs over which they were erected (fig. 5); a problem of stabilisation of rock slopes. Very frequently, in historical cities there are a number of superimposed remains; in fig. 6 a Roman villa of the imperial age found just below a church in a seaside resort near Napoli is reported. The villa had been buried by the Vesuvius eruption of 69 AD, and the church has been built in the Middle Age on a thin tuff layer covering it. It is evident that the solution of any foundation problem for the church needs the use of sophisticated geotechnics.

The famous Leaning Tower of Pisa is depicted in fig. 7; it has been recently stabilised by slightly decreasing its inclination by underexcavation. The inclination of the Big Ben clock tower (fig. 8) caused by the nearby excavation of the tunnels of the Jubilee Line Extension has been controlled by compensation grouting. Both these case histories are reported in details in two contributions in this volume.

The road tunnels excavated by the Romans in the Phlegrean Fields, east of Napoli, are unprecedented and unequalled masterpieces of ancient geotechnical engineering. Some of them have been in regular use till the XIX century! The so called Grotta di Cocceio, linking the Avernus lake to the citadel of Cumae (fig. 9), includes an enormous explosion cavern caused by the firing of explosives stored there during World War II. The cavern is now inhabited by chiroptera (bats) who are a protected species; their

Figure 5.   The village of Terra Murata in the island oProcida. (Left) view from the sea; in the background, at the right of the picture, the Abbey of S. Michele. (Right) bird's eye view.

4

Figure 6. (Left) the Church of S. Maria Assunta at Positano, near Napoli. (Right) the remains of a Roman villa of imperial age found just a couple of meters below the foundation of the church.

Figure 7. The Leaning Tower of Pisa.

Figure 8. The Big Ben Clock Tower.

protection is at present conflicting with the stabilisation of the cavern, epitomizing the complexity and variety of conservation problems.

Many other examples could be quoted, that are omitted here for space reasons.

4 INTEGRITY

The conservation of monuments and historical sites is one of the most challenging problems facing modern civilization. It involves a number of factors belonging to different fields (cultural, humanistic, social, technical, economical and administrative), intertwining in inextricable patterns.

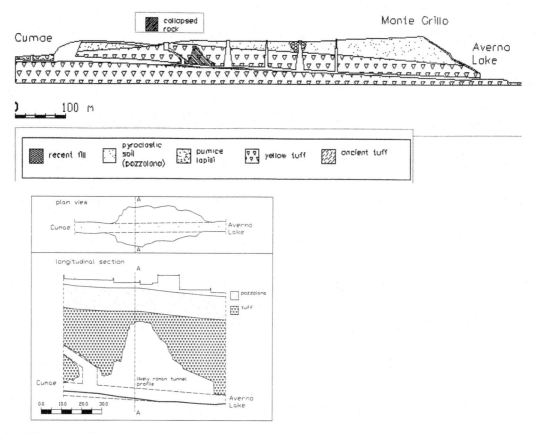

Figure 9. The grotta di Cocceio, a Roman tunnel in the Phlegrean Fields east of Naples. (Above) longitudinal profile of the tunnel. (Left) the explosion cavern in the mid of the tunnel.

From the point of view of an engineer, the peculiarity of this type of intervention is the requirement of respecting the integrity, besides guaranteeing the safe use. The latter requirement is relatively straightforward for a well trained and experienced engineer. The former one, on the contrary, is worth of some discussion.

A prerequisite is the comprehension of the concept of integrity; one soon discovers that it has many facets and is rather elusive. Formal or iconic integrity is the first and most obvious facet; the external aspect, the original form should not be altered by the engineering intervention. Let us imagine an old, experienced civil engineer contemplating Leonardo's Mona Lisa, in the Museum of Louvre in Paris. Let us also imagine that his expert eye discovers that the wall hook to which the picture is hanging is near to collapse, with the risk of the masterpiece falling on the floor and possibly being damaged. What horror! The expert engineer takes a new large nail, plants it in the forehead of the lady and knocks it into the wall (fig. 10). "Now it is safe", he concludes with a legitimate satisfaction; but the visitors of the Museum would probably be much less satisfied.

Unfortunately formal integrity, though being very important, is not all, otherwise a copy would have the same value as the original. Another important facet of integrity is historical integrity; it can be best illustrated by an example, taken from the history of Napoli. The church of Santa Chiara was founded in 1310 by Robert the Wise and Sancia of Mallorca as a double convent for the Poor Clares and Franciscans.

The church exceeded in scale any other church in the kingdom; it loomed over medieval Naples and still presides over the modern city (fig. 11). It was intended not only to host the tombs of the royal family but probably as part of a program to propose a Franciscan alternative to the authority of the papacy, by this time displaced to Avignon. The project of Santa Chiara, initiated soon after the arrival of Robert and Sancia in Napoli, reflected a new trend in the spiritual life of the kingdom, now strongly inclined towards the Franciscan and in particular towards the Spirituals. According to some historians,

6

Figure 10.   Mona Lisa stabilised.

Figure 11.   The Angevin church of Santa Chiara in Napoli.

Santa Chiara was perhaps nothing less than the setting for a brave and doomed attempt to reform the Church.

Before 1943, the medieval interior of Santa Chiara was actually invisible, encased in a sumptuous baroque decoration designed by Domenico Antonio Vaccaro in 1744 and executed by Giovanni del Gaizo and others (fig. 12). It perfectly epitomized the defeat of the Franciscan party and the historical victory of the Rome papacy.

7

Figure 12.   The interior of Santa Chiara in 1942.

Figure 13.   Santa Chiara in August 1943.

Figure 14.   Has the historical integrity been respected?

The original structure underneath the baroque decoration was revealed to modern eyes only on 4 August 1943, after American incendiary bombs caused a fire that burned continuously for 36 hours. The eighteenth century stucco was entirely destroyed and the medieval walls behind severely calcinated (fig. 13). In keeping with the post-war preferences for streamlined design, and principally because of the cost and complexity of re-creating the splendors of the baroque interiors, the church was reconstructed instead to an austere medieval shell, a loss still lamented by most Neapolitans lovers of the Baroque. Has the historical integrity been respected? (fig. 14).

A further facet of the integrity is the material integrity. It is at present acknowledged that the materials, the construction techniques, the structural scheme are original features of the monument as significant as its appearance and history. An arch should not be transformed into a garland of stones hanging from a hidden beam; a direct foundation should not be transformed into a piled one, unless this is the only way to save it, and in any case at the price of a defeat.

Fig. 15 shows one of the solutions proposed in the 1970's to stabilize the leaning tower of Pisa; while completely exhaustive from a merely technical viewpoint, it would have deprived the visitors of the tower of the subtle sensation of risk when climbing the spiral stairs!

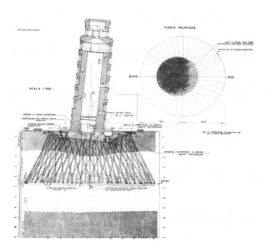

Figure 15.   One of the solution proposed in the 1970's for the stabilization of the Tower of Pisa.

Figure 16.   Pienza.

Finally, we have already mentioned the harmony between the city and the wider landscape that is one of the factors making Italy so important from the point of view of cultural heritage. Fig. 16 shows a quintessentially Italian landscape: Pienza in Tuscany, a small city founded by Pope Pius II in 1462 and still gloriously emerging from the surrounding landscape on top of a hill, in the Senese Val d'Orcia. The balance of countryside and cityscape is so admirably preserved that in 2004 the entire Val d'Orcia was included among the UNESCO sites.

Geotechnical interventions may significantly affect the landscape; this is another important facet of integrity to take into account. The cliff in the island of Procida below Terra Murata mentioned above, for instance, could be protected from the sea action lining it by reinforced concrete, but this would be an irreparable wound to the landscape.

5 SHARED CULTURE

The requirements of safety and use, in the majority of cases the Author has experienced, appear (and often actually are) in conflict with the respect of the iconic, historical and material integrity of the monuments. In almost all countries of the world conservation is looked after by an official trained in Art History or Archaeology. Generally (e.g., this is the case in Italy) he has an absolute control on any action to be undertaken, and imposes constraints and limitations that sometimes appear unreasonable to the engineer. The engineer, in turn, tends to achieve safety by means of solutions which appear unacceptable to the official in charge of conservation, sometimes mechanically applying procedures and regulations conceived for new structures. With a misused word, he tends to cementify.

It is evident that some equilibrium has to be found. Conservation requires on one hand the safeguard of the formal, material and historical integrity of the monument, while on the other its survival and safe fruition. The difficulty of the problem is increased by the lack of a general theory, guiding the behaviour of the various actors involved as Mechanics does with the structural engineer. The lack of unicity of the solution of preservation and conservation problems, vividly exemplified by the case of Santa Chiara, is in particular rather disturbing for us engineers.

It is a deep belief of the author that a satisfactory equilibrium between safety and conservation, between engineers and restorers, may be found only in the development of a shared culture. In the last decades, significant advancements have been actually registered in this direction between the realm of conservation ant that of engineering, and a number of associations, conferences, seminars have contributed to these advancements. ISSMGE is pursuing this goal in different ways; among them, the institution of a Technical Committee on Geotechnical Engineering for the Preservation of Monuments and Historic Sites (TC301, formerly TC29), started many years ago by Jean Kerisel and Arrigo Croce and sponsored by the Italian Geotechnical Society. This volume is intended as a contribution to the effort.

6 NECESSITY AND SUFFICIENCE

Unfortunately, the development of a shared culture of conservation is a necessary, but far from sufficient condition. Immediately after the inclusion among the UNESCO sites, the area around Pienza was involved in a real estate project (fig. 17). The new settlement, the *Casali di Monticchiello*, were advertised as "your new home in a Unesco site". In other words, the UNESCO label, that the Val d'Orcia earned for its preservation, was immediately exploited for commercial reasons.

The eighteen-year-old king Charles of Bourbon, who entered Naples to great celebrations in 1734, inaugurated a new era in the history of the Kingdom which was now once more independent after centuries of being a Spanish viceroyalty. He initiated the digs in Herculaneum (from 1738) and Pompeii (from 1748), which produced an enormous quantity of new antiquities. This situation gave rise to Neapolitan legislation to protect the cultural heritage (1755), expressing the King's "profound regret" over the past export of antiquities from the Kingdom and establishing new rules to prevent it to happen again in the future.

Indeed, when Charles III became King of Spain (1759), in his new capacity he did not issue any provisions to safeguard artifacts there. Had his "profound regret" over the lack of protection for works of

Figure 17.   Commercial exploitation of Val d'Orcia near Pienza.

art in Naples vanished once he got to Madrid? No. In both cases, the monarch had not been writing the legislation personally, but expressing through it the civic and juridical traditions and practice of the place in which he was ruler.

In Italy, there is a sharp contrast between a long history of preservation and its decline over the last few years. The roots of the civic, cultural and juridical aptitude to preservation are to be found in the spirit and tradition of the Italian cities, which, at least from the twelfth century on, had been developing a deeply held and highly sophisticated concept of citizenship, in which the monuments were the basis for civic pride and identity and a sense of belonging which were closely linked to the very idea of a well governed community. In this connection, it is interesting to quote two documents:

i. the Commune of Rome (1162), concerning Trajan's Column (fig.18), states: "In order that the public honor of the City of Rome is preserved, the Column shall never be damaged or knocked down, but must remain as it is for eternity, intact and unspoiled for as long as the world shall exist. Should anyone inflict or attempt to inflict damage on it, they shall be condemned to death and their assets confiscated by the treasury";

ii. the Constitution of the Commune of Siena (1309) says that "those who govern the city must above all ensure its beauty and ornament (fig. 19), which is essential for the delight and amusement of foreigners, but also for the honor and prosperity of the Sienese themselves".

Figure 18.   Rome, Trajan column.

11

Figure 19.   Siena.

Very similar principles are found in hundreds of documents: beauty, decorum, suitability, public honor, the common good or *public benefit*, for which the classical Roman formula *publica utilitas* was often employed. There is a perfect continuity between the conservationist laws of Italy's liberal governments, the two laws passed by Mussolini's regime and finally Article 9 of the Republic's Constitution; this come as a surprise only to those who think in terms of labels and affiliations, and fail to enter into the complexities of the history of ideas. What might be even more surprising is the evident continuity between the conservationist legislation of the Italian states of the *Ancien Régime* (for instance, papal Rome and Bourbon Naples) and the heritage and conservation culture that spread around Europe after the French Revolution. The latter was definitely not a "restoration" of previous laws, but a radical rethinking of the language and rules of the *Ancien Régime* in the light of new guiding principles, such as the concepts of nation, of popular sovereignty and of citizenship, which the events of the French Revolution had changed forever, while giving new meaning to the notion of the "common good" and encapsulating it, among other things, also in historical monuments.

However, this complex system is operating in an increasingly ineffective manner. The devastation of the landscape in Italy has become dramatic. The harmonious relationship between the Italian cities and their countryside, established over many centuries, is giving way to an uncontrolled urban sprawl, which is now home to a large proportion of the population. Although the conservationist laws remain in force and indeed are constantly improved (on paper), "derogations", "exceptions" and even "amnesties" (*condoni*) for the infringement of building regulations are continuously enacted. At the same time, conservation of the cultural heritage is undergoing a deep crisis caused by a lack of human and financial resources.

I have a dream. The international geotechnical community, besides contributing to the development of a shared culture of conservation, plays an active role in the re-establishment of these principles in the common sentiment of people in Italy and all over the world.

ACKNOWLEDGMENT

The author owes a deep debt for their teaching and for stimulating discussions over the years to masters as Arrigo Croce, Ruggiero Jappelli, Jean Kerisel and to friends as Giovanni Calabresi, Salvatore D'Agostino and Salvatore Settis; their inspiration is gratefully acknowledged.

Geotechnics and Heritage – Bilotta, Flora, Lirer & Viggiani (eds)
© 2013 Taylor & Francis Group, London, ISBN 978-1-138-00054-4

Geotechnical issues of the Athenian Acropolis

D. Egglezos
NTUA—TC301 Member, Athens, Greece

M. Ioannidou
Emerita of Acropolis Restoration Service (YSMA), Athens, Greece

D. Moullou
Hellenic Ministry of Education and Religious Affairs, Culture and Sports, Greece

I. Kalogeras
National Observatory of Athens, Institute of Geodynamics, Athens, Greece

ABSTRACT: The monuments of the Acropolis survived until today, through centuries of perils and changes in use and in form, wounded but still standing. Their continuous exposure in the action of damaging factors (natural or man made) during their long history has provoked major or minor failures, related to geometry and/or mechanical strength of their structure and materials. Those damages had to be confronted and thus an integrated restoration project began in 1975 and is continued until today. This paper, giving an insight to the restoration programme, focuses on the geotechnical issues of the monumental complex. It presents the geotechnical data collected as well as the most significant studies and interventions already conducted. Moreover, the seismotectonic regime of the area is presented with a commentary review of literature for the historical and recent seismicity of the broader area, as well as a quotation of earthquake effects on Acropolis hill and monuments. Specific macroseismic observations of strong earthquakes of different source properties are mentioned. Finally, a brief description of the strong motion array deployed on the Acropolis hill and some conclusions from the accelerographic records are presented.

1   INTRODUCTION—THE ACROPOLIS MONUMENTS

The Acropolis of Athens is the most striking and complete ancient Greek monumental complex still existing in our times; an architectural treasure that belongs not only to the Greek patrimony but also to the worldwide cultural heritage (Fig. 1). It is situated on a medium high hill (altitude 157 m) that rises in the basin of Athens. Its overall dimensions are approximately 170 × 350 m. The hill is rocky and sheer on all sides except for the west, and has an extensive—nearly—flat top covering an area of about 30.000 m².

Strong fortification walls have surrounded the summit of the Acropolis for more than 3.300 years. The first fortification wall was built during the 13th century BC (Fig. 2), enclosing the residence of the local Mycenaean ruler. In the 8th century BC, the Acropolis gradually acquired a religious character with the establishment of the cult of Athena, the city's patron goddess. The sanctuary reached its peak in the archaic period (mid 6th century-early 5th century BC).

In the 5th century BC, the Athenians, empowered from their victory over the Persians, built a new circuit wall and under the leadership of the great statesman Perikles they carried out an ambitious building programme comprising a large number of monuments that transformed the rocky hill into a unique complex, which proclaimed the ascendancy of classical Greek thought and art. These monuments include: the Parthenon, a temple of Doric order, of particularly large proportions, that dominates the summit of the Acropolis rock (Fig. 3); the Erechtheion, a temple of Ionic order, of curious shape, with rich sculptural decoration and the emblematic feature of the Porch of the Caryatides

Figure 1.   The Acropolis of Athens.

Figure 2.   Part of the Mycenaean fortification wall preserved on the Acropolis hill.

Figure 3. The Parthenon (right) and the Erechtheion (left).

(Fig. 3); the Propylaia, the monumental entrance building to the sanctuary, famous for its impressive coffered ceilings (Fig. 4); the temple of Athena Nike, a small temple of the Ionic order (Fig. 5), notable for its elegance and grace (Brouskari 1997, Ioannidou et al., 2008). These buildings, characterized by ingenious planning and flawless construction, were built with dry masonry consisting of white pentelic marble. Only the foundations were of poros stone.

Smaller buildings, a plethora of votive offerings and inscriptions, fill the picture of a living and active sanctuary that remained as such until the 4th century A.D. From the end of the 3rd century A.D., however, the west side of the Acropolis is fortified (Ioannidou et al., 2008). With the establishment of the new religion—Christianity—the process of decay began. The sanctuary was stripped of its artistic wealth, some of the smaller buildings were destroyed and significant changes were made to the larger ones that survived: the Parthenon was converted into a church of the Holy Wisdom, popularly known as the church of the Virgin Athiniotissa, the Erechtheion was transformed into a Christian basilica dedicated to the Virgin. A church was built in the south-west wing of the Propylaia in the 6th century A.D., and in the 10th century the main part of the monument was converted into a church of the Archangels (Brouskari 1997).

Figure 4. The coffered ceiling of the Propylaia. Restored.

Figure 5. The temple of Athena Nike after its restoration.

Figure 6. The Acropolis in the mid-15th century.

Figure 7. General view of the Acropolis at the begining of the 19th century. E. Dodwell (1804–1806).

During the Latin domination (1204–1458) extensive fortification works were carried out (Fig. 6) and the Propylaia was transformed into a fortified residence for the ruler of Athens. During the Ottoman domination (1438–1833) the Parthenon becomes a mosque and the Erechtheion the residence of the local ruler, the disdar (until the 18th century). Parthenon was severely damaged in 1687, when it was bombarded by the Venetians. The explosion of the gunpowder stored in the building transformed it to a ruin (Brouskari 1997) (Fig. 7).

After the end of the Greek War for Independence, during which the Acropolis was found in the centre of the events, and the creation of an independent state (1833), the Acropolis was proclaimed an archaeological site (Ioannidou et al., 2008).

Since then works began on cleaning the ruins and restoring the monuments that have survived for almost twenty-five centuries through wars, explosions, bombardments, fires, earthquakes, thefts and interventions as well as alterations connected with different usage.

The extensive restorations carried out, throughout the 19th and at the beginning of 20th century, together with the still ongoing restoration programme that began in 1975, gave the Acropolis monuments the appearance they have today.

2 GEOTECHNICAL DESIGN AND PATHOLOGY OF THE ACROPOLIS MONUMENTS

2.1 *Geology of the Acropolis hill*

The Acropolis hill is a trapezoidal-shaped block of Late Cretaceous grey limestone (thick platy or unbedded) resting on the marls and sandstones of the Athens Schist rock series, encountered either as marly schist, or sandstone marl (Fig. 8).

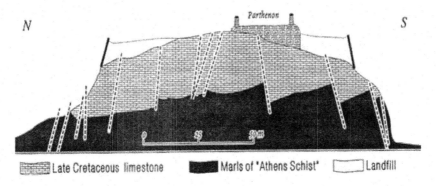

Figure 8. Geological section of the rockhill (N-S direction). From Higgins and Higgins 1996.

Between the limestone mass and the underlying bedrock a rather thin layer of conglomerate is interjected. Locally, in the upper surface of the bedrock conglomerate horizons are met. Superficially the bedrock is covered at places with an eluvial soil cover of small thickness, coming from the weathering of the schist bedrock.

The grey limestone is well exposed on the top of the hill. It has closely spaced joints and some of the older fissures have been filled with red marl or coarse calcite crystals. The top of the hill has been leveled with artificial fill up to 17 m thick which is retained by the Circuit Wall (see paragraph 2.3). The artificial fill that forms the plateau where the monuments stand has a varying composition. This fact is mainly attributed to the different historic stages of backfill construction.

The prevailing theory for the creation of the Acropolis rock hill considers an initial unified large formation of limestone mass including all the neighboring hills (Areios Pagos, Filopappos hill, Pnyx etc). As a result of the strong erosive action of the environment, the initial mass has been divided into several hill formations that exist nowadays.

The rock mass quality shows significant local differentiations. The summit and the inner part of the hill consist mainly from compact to thick platy limestone and show a rather low degree of weathering with sparse closed discontinuities and favorable tectonic characteristics. On the other hand the slopes show clearly a higher degree of weathering (leading to systematically jointed rock mass). This is especially clear in the case of the east and the north slope, where the intense karstic effects, the usually intensely fractured rock mass, the unfavorable orientation of the discontinuities and the increased action of the environmental agents, lead to the development of local instabilities and rock falling as well as to the formation of caves and clefts. Such caves can be seen in the slopes of the hill. According to Higgins and Higgins (1996), other caves, probably mostly choked with debris, exist in the interior of the hill.

The discontinuities of the permeable limestone formation allow for a regular drainage of the hill. The water from the rainfalls, seeping into the faults and fractures on the upper part of the Acropolis, discharges in characteristic springs at the hill base, since the underlying schist bedrock is practically impermeable.

As far as the tectonics of the rock are concerned, 23 micro tectonic diagrams were performed (Andronopoulos and Koukis, 1976): 12 around the slopes and 11 at the summit of the hill. The most unfavorable situations are met in the eastern slope and the eastern crown (5 locations), in the northern slope (3 locations), while in the south slope only 1 location presents unfavorable tectonic characteristics.

The Acropolis hill was not intensely defaced with quarries. Limestone from several of the adjacent hills including the Hill of the Nymphs was used for many of its buildings. Dolomite and marly limestone that came from Piraeus was used for many buildings' foundations and the Circuit Wall. Marble from Mt. Penteli was used for the construction of all the great buildings of the Periclean project. Besides these materials, limestone and conglomerate from Kara, near the base of Mt Hymettos, were also extensively used (Higgins and Higgins 1996).

2.2 The foundations of the existing monumental structures

As it was previously mentioned (paragraph 1) a plethora of buildings—small or large—once stood on the Acropolis hill throughout its long history. From those buildings only four remain standing: the Parthenon, the Erechtheion, the Temple of Athena Nike and the Propylaia. However, the foundations of other ancient structures are still—clearly—visible on the Acropolis plateau. Among them are the *Old temple of Athena*, a large Doric temple of the 6th century B.C. that was destroyed by the Persians in 480 B.C., *the Arrephorion*, a small cult building of the 5th century B.C. where the mystery ritual of the Arrephoria took place, the *Chalkotheke*, a rectangular building of the 5th—early 4th century B.C. that housed the metal offerings and dedications to the patron goddess, the *sanctuary of Artemis Brauroneia*, an irregular Π shaped building (in the form of a stoa with two projecting wings), founded by the tyrant Peisistratos in the 6th century B.C., *the sanctuary of Zeus Polieus*, where the ceremony of Diipolia was held, *the temple of Rome and Augustus*, a circular monopteral temple of the late 1st century B.C. and the *monument of Agrippa*, a high pedestal that supported a bronze quadriga (with the statue of Agrippa in it), probably originally erected by Eumenes II in 2nd century B.C. with the statues of himself and his brother Attalos II in the quadriga.

Most of the above buildings were founded directly on the natural rock. But in some cases, mainly due to the irregular geometry of the rock at the location selected for the erection of each monumental structure, typical foundations from poros stones had to be constructed. The cases with foundations of geotechnical interest are the Parthenon, the Propylaia, the temple of Athena Nike, the Arrephorion and the monument of Agrippa.

2.2.1 The Parthenon

The Parthenon was preceded on the same site by an earlier temple, of equally large proportions, conventionally known as the Pre-Parthenon. The Pre-Parthenon's foundations (stereobate) were used also for the foundations of the Parthenon of the Periklean era (Brouskari 1997, 153). (Fig. 9a, b).

The preserved temple stands on the top of the rock. Only three quarters of the building possess foundations. The material is poros ashlars quarried especially for the purpose (Bundgaard 1976, 54). For the necessary elevations see Fig. 10.

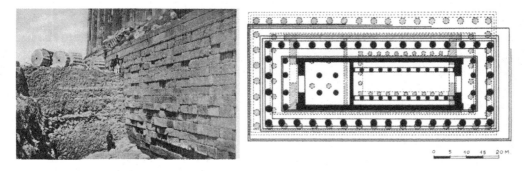

Figure 9.   Parthenon foundation a) South side after Cavvadias and Kawerau 1907 and b) plan. (Periklean in red, pre- Periklean in black). After Brouskari 1997, 152.

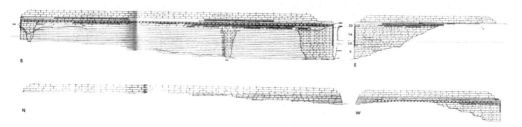

Figure 10.   Elevations of the Parthenon's stereobate. (After Bundgaard 1976, t.H)

"The foundation attains its greatest depth in the south-east corner and the twenty adjacent meters of the south side. Here it has 22 courses. The bottom course consists of a single row of stretchers, the second of a single row of headers. The third has four rows of stretchers, the fourth two rows of headers and so on up to and including course 19, the fourth course from the top. The three top courses follow the usual principle, the third from the top having headers in both places, the second stretchers and the first again headers. ... The ashlars in the three top courses fit tightly all the way round. In these three courses the jointing is executed with the same precision as in a marble wall." (Bundgaard 1976, 54).

The interior of the foundation is not accessible. Most scholars believe that it is probably a complete block of stone. However, the opinion that the foundations consist of separate walls has been expressed (for a discussion see Bundgaard 1976, 57–60).

As far as the geotechnical design is concerned, this type of foundation based upon the natural limestone rock allows for high values of bearing capacity and minimal—elastic—subsidence (settlement) from the action of the structural loads.

This design creates a firm base for carrying safely the vertical superstructure loads. Additionally, the dry masonry structure assures a relatively flexible foundation base capable to absorb high energy arising from the propagation of seismic waves. This conclusion is based on the fact that the acceleration recordings upon the crown of the Parthenon foundation are systematically lower than the recordings upon the free surface of the rockhill (Kalogeras and Egglezos 2013).

2.2.2 The Propylaia

The Propylaia consist of a main building (24 × 18.20 m) and two wings on the north-west and south-west sides. The east section of the main building, following the ridge of the rock, on which it stands, was at a higher level than the west. The north-west wing, called the Pinakotheke (building with paintings), is situated lower than the main building, standing on a high podium. The Pinakotheke consists of a rectangular room (10.76 × 8.97 m) and a porch with three Doric columns in antis facing south (Figs. 11–12). The south–west wing, having three Doric columns in antis, is placed symmetrically regarding the north-west wing, it was of the same height and faced north. However, its room was smaller (8.97 × 5.23 m) and had no west wall. The west wall was replaced by a double opening with a central pillar partition and a free standing double pillar at its north-west corner (Brouskari 1997, 57–70). The foundations of the south-west wing stand upon the bedrock.

The foundation of the north-west building is built with large limestone blocks onto the leveled rock, consisting of isolated walls with backfilling. At the western side and the west half of the north side it is perceived as a continuation of the Acropolis Circuit Wall and at the same time as a particular architectural form, i.e. as the podium of the Pinakotheke (Korres 1995, 466).

At the south side, the visible part of this infrastructure is luxurious, as in the corresponding part of the south-west wing. The outer blocks are made of marble and create a special form of wall. This wall is both the marble pedestal of the crepis of the front side and the side limit of the upward courtyard of the Propylaia. This wall, having undergone modern repairs (1854), has a tiered structure and an oblique direction. In antiquity its height was 6 m and its width about 1.3 m. Its inner inclination ratio was 1:10 (Korres 1995, 466–470).

The ancient backfilling of the podium was fairly dense and prevented the structure from major failures. However, the demolition of the medieval structural additions (e.g. the Frankish doms) and the intensive archaeological excavations that replaced the fill with loose earth, during the 19th century interventions (see paragraph 1), contributed to the gathering of rain water inside the poros foundation and the increase of horizontal earth pressures.

Today the Pinakotheke suffers from structural deformations e.g. 3 cm general settlement and small horizontal deformation (inward or outward locally) of the North Wall, general settlement from 3 to 7 cm, outward displacement, barreling and inclination of the west wall etc.

In 1955, in order to secure the western part of the north-west wing, a concrete floor was constructed, the excavation fill was removed and wells filled with reinforced concrete were created. Recently the floor was repaired with carbon fibers (FRP).

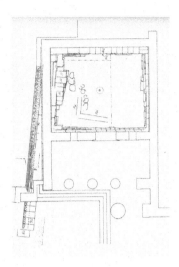

Figure 11.   Pinakotheke and the Agrippa pedestal.   Figure 12.   Plan view of the Pinakotheke (M. Korres).

2.2.3   *The temple of Athena Nike*

The temple of Athena Nike lies on the top of the bastion at the south-west edge of the Acropolis. The core of the bastion is a strong tower of the "Cyclopian" wall, dating back to the Mycenaean era. Inside the classical bastion and under the classical temple exist, among others, the remains of a small poros temple, in the form of a simple oikos (Fig. 13–14). (For a full description of the pre- classical deposits see indicatively Brouskari 1997, 73–77). The Mycenaean tower was coated with isodomic masonry of poros stones and its internal part was filled with soil.

The temple was demolished by the Ottomans in AD 1686 and its members were used to create a defense wall that had been built between the Nike bastion and the Pinakotheke of the Propylaia. The members of the temple were discovered and identified during the 19th century demolition and cleaning works. It has been restored three times (1835–1844, 1935–1940, 2003–2010). During the recent restoration, flooring in the form of a stainless steel grid over the basement space beneath the temple of Athena Nike was installed (Fig. 15).

2.2.4   *The Arrephorion*

The preserved parts of the building include the foundation walls, which form two underground halls. The dimensions of the central hall (A 1) are 8.30 × 4.30 sq. m and the maximum height of the foundation 4.90 m. The smaller hall, that is the porch (A 2), measures 8.30 × 2.30 sq. m and the maximum height of the foundation wall is 3.00 m. The foundation walls, which are rectangular blocks of poros-stone (brittle marl limestone from Piraeus), range between 1.40 and 2.00 m in thickness. (Fig. 16).

2.2.5   *The monument of Agrippa*

Today the only remains of the monument of Agrippa are the pedestal that supported the bronze quadriga. The pedestal consists of a high, rectangular foundation (4.50 m high and measuring 3.31 × 3.80 m) constructed of conglomerate and poros and a main section of Hymettian marble, 8.91 m tall that tapers slightly towards the top. The projecting cornice and base are of white Pentelic marble. The main section is built in pseudo-isodomic masonry (Brouskari 1997, 53) (Fig. 11).

The foundation was stepped to agree with the inclination of the ramp served in antiquity for the approach to the sanctuary (Stevens 1946, 90–91).

The structural failures of the monument include fractures of the foundation stone blocks and intense rotation of the main section, mainly attributed to differential settlements due to the erosion of the conglomerate and poros stones of the foundation. Possibly seismic actions may have worsened the monument's state.

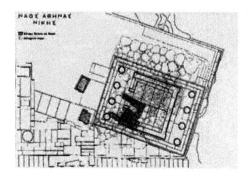

Figure 13. Plan view of the Nike bastion (V. Douras).

Figure 14. The Nike bastion after the dismantling of the temple.

Figure 15. Athena Nike. The stainless steel grid.

Figure 16. Plan view of the Arrephorion.

2.3 The Circuit Wall

Among the visible monuments on the Acropolis hill, only the Circuit Wall serves a pure geotechnical purpose, since it functions as a typical gravity wall retaining the backfill that forms the plateau of the Acropolis. Therefore it is of particular geotechnical interest.

The Circuit Wall is dated in the classical era (5th century B.C.). It has a polygonal perimeter of about 750 m and height ranging from 4 to 18 m. Its construction does not follow a single plan. It was constructed in sections, in different phases and with different construction forms.

2.3.1 The North Wall

The North Wall, generally known as the "Themistokleian Wall", is considered to be one of the first structures erected by the Athenians on the Acropolis hill after the Persian Wars (479 B.C). Being 320 m long and 4–6 meters high, it is built with dry masonry. Unlike that on the south, the North Wall does not show—from cursory glance- continuity of form or of construction (Korres 2002, 179). The isodomic construction is interrupted in various points by the reuse of marble and poros stone architectural members of earlier buildings that stood on the summit of the Acropolis before the Persian invasion, including 26 half-finished unfluted marble column drums from the Older Parthenon (Fig. 17) and the entablature (architrave blocks, triglyphs, metopes and cornice blocks) of the "Old Temple of Athena", laid in a continuous row as crown blocks on the wall (Fig. 18). Most of the ancient construction is still visible. Later repairs, some of them filled in with rubble masonry, can be recognized.

The North Wall is a typical retaining gravity wall and its pathology is similar to that of the south (see par. 2.3.2). As far as the geometry is concerned, the typical section of the North Wall is significantly shorter and thinner than the south one (retains less earth fill). From a static point of view, the presence of reused architectural members of high quality normally enforces the structural performance of the

Figure 17. View of the North Wall. Part.

Figure 18. Section of the North Wall.

wall. However, there are areas where the geometry of the members creates unfavorable stability conditions (see Egglezos and Moullou 2011).

2.3.2 *The South Wall*

The South Wall was built by the Athenian leader Kimon after his successful military campaign of 467 BC in Asia Minor. The wall measuring 295 m long, has a 165 m eastern leg and a 130 m western one. The two legs meet at an angle of 152.5 degrees. Its height ranges from 10 to 18 meters. The wall is founded on the natural carved rock and consists of orthogonal stone blocks with typical dimensions: 1.30 m × 0.65 m × 0.50 m (L × B × H) (Fig. 19).

The upper part of the wall presents two characteristic typologies. The SE part has a crown consisting of a huge orthogonal platform much thicker than the underneath section, while the SW part has a section with a decreasing width from bottom to top (Fig. 20). The limit between these two formations practically coincides to the middle of the Parthenon's south side. It is suggested that the upper part of the SE wall with the wide summit was built under Perikles (Korres 2004, 278).

During the history of the wall, the unfavorable action of natural agents and the extreme (permanent or temporary) loading actions (mainly attributed to the enormous pressure of the earth fill behind it) led to structural failures of either geometrical (displacements, rotations, barreling etc) or mechanical (fractures, collapses etc) nature. These failures forced the need for systematic reinforcement of the wall during the medieval and the ottoman era. Strong buttresses (8 narrow and 3 wide), thick revetments (covering almost the whole western leg), extensive later repairs with rubble masonry and locally a new mortar facing (SE area), hide the original masonry from view, transforming the South Wall to a hybrid structure consisting of the inner ancient wall and a newer face—not connected practically with the ancient part- of masonry with mortar.

2.3.3 *The East Wall*

The East Wall has a total length of about 100 m and its height ranges from 10 to 18 m (Fig. 40). It seems to be a continuation of the South Wall since the wide summit (observed in the SE part of the South Wall) continues to the southern part of the East Wall. It has been suggested that the northern part of the wall was not built before the 3rd century A.D. (Korres 2004, 281).

A large part of the East Wall collapsed in an earthquake in 1705 and it was rebuilt as a typical masonry construction with mortar in the mid 18th century (Korres 2004, 281).

2.3.4 *The artificial fill*

Due to the morphology of the hill the soil filling prism is much thicker in the south area of the hill, in comparison to the north. The fill is associated with the relevant constructional phases of the Circuit Wall and the history of the Acropolis hill in general. The ancient fill, which besides common cobbles comprised many architectural members, sculptures and debris from the plethora of buildings and votive offerings that once stood on the hill, was almost completely removed, during the intensive

Figure 19.   The South Wall.

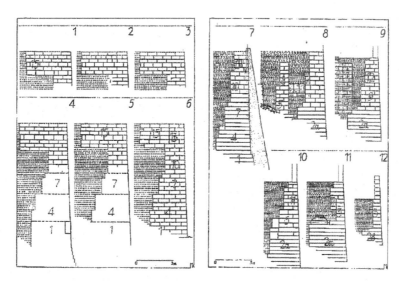

Figure 20.   Characteristic sections of the South Wall. South-east leg sections: no. 1–6, south-west leg: no. 7–12 (Korres, 2004).

excavations of the 19th century. The exact layering of the strata is described in the excavations' report (Cavvadias and Kawerau 1907).

The quality of the ancient fill can be deduced from the photographs taken during the excavations of the late 19th century, where a vertical slope of almost 10 m height stands stable (Fig. 9a). Consequently both cohesion and elaborated compaction may be attributed to the ancient fill.

The same material was re-used for the backfill after the excavations. Of course, the current state of compaction of the fill should be considered much looser than the ancient one.

Although the composition of the fill varies greatly from site to site a general pattern could be considered for geotechnical purposes. From a geotechnical point of view the fill can be classified as compacted gravel (GC-GW) to rockfill. However, it should be mentioned that the degree of compaction varies locally. For example during the 2007 excavation of the Arrephorion the backfill consisted of rockfill, whereas the 2010 excavation in the west side of Parthenon revealed well compacted clay gravel with cobbles.

2.3.5 *Loading actions on the Circuit Wall and "pathology"*
The loading actions on the Circuit Wall can be summarized as follows:

- Self weight
- Earth pressure (mainly governed by the nature and the compaction of the artificial filling, the excavations, the refilling compaction, the rainwater effect, the live loads on the surface and the seismic motion)

22

- Hydraulic action (increase of unit weight due to saturation, smoothing of the stone blocks joints, hydrostatic pressure behind the low permeability parts of the wall, natural degradation of the stone blocks with loss of mass)
- Quasi—permanent loads as construction of battlements, buildings etc
- Military action (bombardment of the wall etc)
- Seismic action.

As a result of the above factors, the monument shows in places a series of typical problems of a structural nature, which compose its "pathology".

Generally speaking, the structural problems can be divided into three basic categories:

a. those of a geometrical nature, which are connected with change in the form of the structure, without failure of the structural members, such as displacement, rotations, barreling, inclinations relating to the vertical axis, etc. The failure of this nature is mainly attributed to the huge earth pressures and to the effect of seismic events.
b. those of a mechanical nature, connected with strains and failures of the structural members without change to the form, such as cracks, breaks, detachments of stones, weathering, etc. The above structural problems have created areas of weakness locally, and, to begin with, they raise concern as to the capability of the structure to respond satisfactorily to increased transient loading (such as hydrostatic pressures from heavy rainfall, earthquake, etc.). The effect of this failure is mainly attributed to the great compressive and/or tensile stresses developing on the wall mass as a result of various loadings.
c. those attributed to the unfavorable influence of natural physicochemical agents. This kind of failure comprises cracks, loss of mass, soilification of the stone, mortar leach and so on.

2.3.6 *Evaluation of the ancient design in relation to contemporary geotechnical knowledge and seismic risk analysis*

2.3.6.1 *Qualitative assessment of the Circuit Wall design*

The structural system of the Wall is typical dry masonry consisting of stone blocks. These blocks are cooperating through frictional forces. The advantage of such a structure comprises the facility in construction, the low requirement for specialized work and the easy adaptation to the natural environment.

Concerning the mechanical performance of this kind of structure, the following apply to the Wall's structural design: the compressive strength is excellent (that of the stone material), the shear strength is offered from the frictional forces between the stone blocks interfaces, and a quasi tensile strength may be considered arising from the gravity (vertically) and the friction (horizontally).

The drainage of the Wall is generally satisfactory and the draining of the rainwater is achieved through the stone block joints. However since a typical drainage soil layer behind the Wall never has been put, the composition of the backfill and its permeability locally, may greatly influence the drainage capacity of the Wall.

The Wall is founded upon the elaborated surface of the limestone rockhill slope. As a rule the foundation upon this geotechnical unit offers sufficient bearing capacity. Issues of the rock slope stability will be addressed in a greater extent in paragraph 5.2.1.

2.3.6.2 *Assessment of the dimensioning*

According o the empirical rule for typical gravity walls a sufficient design is assured for a base to height ratio $0.35 < B/H < 0.70$.

Concerning the South Wall, the following apply to the wall geometry: in the areas with no buttresses $0.33 < B/H < 0.50$, while in the areas with buttresses $0.40 < B/H < 0.60$. In any case the design is acceptable (especially if the quality of the backfilling is also considered).

Concerning the North Wall a $0.25 < B/H < 0.35$ ratio applies, which is a marginally acceptable design. This remark is in good agreement with the state of preservation of the North Wall (bad to problematic in many areas) and the extended structural failures as concluded from the repairs in significant parts of the Wall.

The backfill is a well compacted (at least in the ancient times) gravel (GC-GW) applied in layers of about 0.50 m height. Of course after the replacement of the fill with the completion of the 19th century excavation a rather lower compaction may have been attained (absence of relevant evidence on this issue).

2.3.6.3 *Empirical assessment of the seismic risk of the Circuit Wall*

An initial assessment of the expected seismic damages in monumental structures may be derived through vulnerability curves. The typical risk diagrams show the possibility that various levels of damages (light, medium, heavy) take place as a result of the seismic acceleration and the geometry of the structure. The vulnerability of the Circuit Wall is assessed using relevant curves for gravity walls (Kakderi and Pitilakis, 2010). For the Wall case the term damage is quantified as the ratio of the Wall crown drift to the Wall height (a ratio < 1.5% refers to a light damage, a ratio >1.5% and <5% to medium damage and a ratio >5% to heavy damages). The expected PGA's on the site (according to EC-8 provisions, and taking into account the soil and the topographic effect) range from 0.28 to 0.40 g. From the diagram results a high possibility for light damages of the overall wall structure. Additionally, it is shown that the North Wall is more vulnerable than the south one. In any case the possibility for heavy damages is rather low. This conclusion is consistent with the macroscopic appearance of the monument (and the historically recorded failures as well).

2.3.6.4 *Synopsis of the ancient geotechnical design of the Circuit Wall*

The design of the Circuit Wall, although empirical, is an excellent example of the accumulated experience of ancient Greeks in the design of retaining structures. In fact, this is easily proven from the long-term life of the Wall (more than 2500 years of incessant functioning).

From the qualitative assessment of the Wall's design derive the following conclusions:

- The design for static conditions is considered satisfactory
- The outer revetment and the buttresses contribute to an increased stability of the Wall
- The effect of the upper platform in the SE part of the Wall is beneficial
- The seismic resistance of the North Wall against overturning is marginally sufficient (and locally insufficient).

Of course, additional intervention measures can contribute to an increased safety and protection of the Wall's structure. Specifically, the South Wall—especially after the stabilizing interventions from the medieval age to the mid 20th century (buttresses, revetment)- seems to be in a satisfactory condition. However, the following issues need to be taken into account:

- The ensuring of a sufficient drainage behind the Wall
- The achievement of a sufficient cooperation of the ancient structure and the outer posterior masonry
- The restitution of the decayed stone blocks.

As far as the North Wall is concerned it has to be mentioned that there is marginal to insufficient safety in the case of a strong seismic event.

The basic conclusion is that an overall and rational plan of intervention is needed for preserving the monument.

3 STRUCTURAL RESTORATION OF THE ACROPOLIS MONUMENTS

3.1 *Earlier and contemporary anastelosis interventions of the Acropolis monuments*

As it was mentioned before (paragraph 1) throughout the 19th century and the first half of the 20th century, to the time of the World War II, the Acropolis monuments began to undergo successive interventions with the main purpose to re- establish their form in Classical Times.

The intervention that most affected the appearance of the Acropolis monuments was the restoration project directed by the civil engineer Nikolaos Balanos (1898–1939). Although these neither respected nor adhered to the structure of the monuments (which, in fact, they altered), visually they may be considered successful, for he managed to retain the character of the monuments as ruins, by using to a great extent ancient material, adding very little that was new. Yet, from a technical standpoint, the interventions were catastrophic. Applying the technology of the time in an inappropriate way, Balanos incorporated iron reinforcements, some large, some small within the architectural members of the classical monuments and encased them in cement, following the contemporary belief that this would counter the problem of their corrosion.

During the years after the World War II, the rusting of these reinforcements was to cause breakage and disintegration of the monuments, making the new restoration inevitable. The new intervention, moreover, had to take account of the faulty static efficiency of the monuments because of their ruined condition and the structural problems of the Acropolis Rock and the Circuit Wall, monuments themselves bearing the traces of a long history.

The establishment of the "Committee for the Preservation of the Acropolis Monuments" (E.S.M.A.) in 1975 inaugurated a new epoch in the approach to restoring the Acropolis monuments. The predominant perception was that the complex matter of restoring the monuments required the collaboration of scholars from many fields, archaeologists, architects, civil engineers and chemical engineers. Thus, to the fullest extent, the desire for an interdisciplinary approach was applied to the restoration of the monuments as formulated in the "Charter of Venice", the internationally accepted theoretical framework that governs interventions on ancient and historical architectural monuments. This committee of interdisciplinary specialists, first of its kind in Greece, undertook at that time to study the serious structural problems and the problems of surface erosion of the monuments and to direct interventions on the monuments of a rescue nature. The year 2000, with the establishment of the Acropolis Restoration Service (Y.S.M.A.) of the Ministry of Culture, saw the beginning of a new phase of anastelosis interventions on the Acropolis comprising an extensive restoration programme on the monuments carried out according to the principles and methodology that had been instituted during the preceding years.

3.2 *Structure and pathology of the Acropolis Monuments*

The monuments of the Acropolis are built of worked stones in the form of rectangular blocks or drums, without mortar, joined to each other with metal clamps and dowels. Although constructed of separate architectural members, their "dry masonry" joining has been done so accurately that in some cases the joints are imperceptible, giving the impression that the construction is continuous (Fig. 5). The purpose of the clamps is to relieve tensile and shear forces when the friction restraint between the members is overtaken. The joining elements (horizontal- clamps or vertical to the layers of stones- dowels) are made of iron, placed in specially cut tenon-holes or sockets and sheathed with cast lead (Fig. 21). The clamps connect members in the same horizontal series and absorb chiefly the tensional forces, while the dowels connect members of successive courses and withstand shear forces. These connecting elements assure the total resistance of the construction, especially against seismic load or deformation due to various other disturbances (violent shifting or displacements, foundation yielding, etc.). The basic purpose of the lead sheathing, however, is to protect the iron of the joining elements from rusting by shielding them from the atmosphere. Apart from that, the lead sheathing insures mechanical continuity between clamp/ dowel and stone and contributes, as a pliable material to the absorption of a part of the shock movement and energy of an earthquake.

The dimensioning of the structural elements of the monuments was determined on the basis of morphological criteria, without any preliminary structural calculation in the modern sense of the term. Even so, the excellent structural performance of the monuments in all the natural actions throughout the 2,500 years of their history shows that the mechanical features of the material they used were known to them and that the structural function of the bearing system had been estimated very close to the mark.

A feature of the Acropolis monuments, as structural systems, is that they fulfill today's requirements for an anti seismic design; they are characterized for their clearly defined structural function, for the regularity of their plan, for the symmetrical arrangement of their bearing elements and mass. The

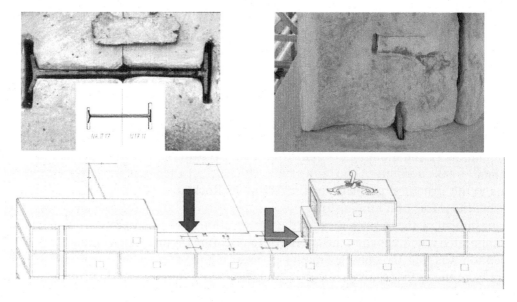

Figure 21. The way of construction of a marble wall with clamps and dowels.

great rigidity of the walls together with the diaphragm function of the ceilings and roofs by means of friction, add to the resistance of the building to horizontal stress. Finally, the founding of the monuments for the most part on solid rock and the good quality of construction of the foundations favor their good anti- seismic behavior. It is notable that a great part of the damage we are obliged to face today is due not to the forces of nature, but to the activity of man.

The failure of earlier interventions on the Acropolis monuments is due chiefly, as we have already mentioned, to the use of ordinary iron for the connecting elements of the fragmentary architectural members or for strengthening other members. Rusting and expansion of the iron elements caused the marble to break and architectural members to shift. The molten lead or cement plaster used in these earlier interventions to fill in gaps was not up to the excellent standard of ancient lead casting, while the hermetic connection of the architectural members in the ancient phase was not achieved in the restoration (Fig. 22). Apart from the rusting of the iron elements of the previous restoration, structural problems are due to other reasons caused by human interventions such as explosions, fires, bombardments, which led to the ruined condition of the monuments, and to physical reasons, as the destruction and displacements caused by earthquake.

In future, the only strong mechanical strains are expected to be seismic, since we certainly hope that there will no longer be damage inflicted by mankind. It is therefore imperative to evaluate the efficiency of the monuments in seismic activity, taking into account the damage they have suffered through their long history and with this evaluation as a basis to make the necessary interventions (Ioannidou 2006, 2007). The analysis of the seismic behaviour of the monuments in their original condition, in any case cannot in itself yield conclusions applicable to today's interventions.

Although recent years have seen great progress internationally in the protection of modern buildings from earthquake, a realistic method of evaluating the behaviour of the classical monuments under seismic load has not yet been developed. The structure of the monuments is based on the perfect contact of the joints. The basic characteristic feature of the structure of the monuments, i.e. the presence of joints between the wall blocks, means that only compression and shear stresses are assumed, and not tensile. This results in a strongly non-linear behaviour of the structure. From the quantitative aspect, controlling the composite movement of the hundreds of members of a classical monument under seismic load is an extremely difficult problem. Apart from the number of joints, the problem is further complicated by cracks, deformations, shifts and failure of joining elements. On the basis of scientific knowledge today, it is practically impossible to make a reliable model simulating seismic load so as to design the interventions needed.

3.3 *General methodology of the intervention*

In cases where the damage is critical, dismantling of areas that have serious problems is inevitable. The dismantling is extended beyond the restored parts of the monuments into areas that have undamaged places here and there if serious damage has extended into these general areas as well. The rusted joining elements are removed and the filling material (cement plaster in the restored areas or, rarely, lead) taken out.

White cement and titanium reinforcements are used for the structural restoration of the stones. The reinforcements are threaded titanium rods which are inserted into holes in the marble mass and secured by an inorganic plaster made of white cement (Fig. 23) (Zambas et al., 1986). The holes do not penetrate to the outer surfaces of the architectural members nor do they reach the coarsely worked interior surfaces, so that they are not visible. Fragments that do not belong together, that is that do not come from the same architectural member, are never joined together.

Where considered necessary, the missing parts of members are added in new marble so that they can regain their original structural self-efficiency. After it has been made, the additional piece is joined to the ancient marble likewise with a titanium rod and cement mortar (Vardoulakis et al., 1995, 2002).

When the architectural members have been repaired, they are set again in their original positions or in positions where they had been placed during the previous restoration (Fig. 23). They are connected with titanium clamps and dowels which are secured in the ancient clamp and dowel cuttings with white cement mortar. During the re-setting of members, geometrical deformations of the area that was dismantled are partially corrected to the extent allowed by the remaining distortions of the members that were not dismantled, in order to achieve as much as possible the original form.

3.4 *Principles of intervention for structural restoration*

The Venice Charter forms the framework of principles accepted internationally in which are codified the requirements necessary for the restoration of monuments. Among them the principles are: the interdisciplinary approach and the collaboration of scholars of different specialities, demanded by Article 2 of the Charter, the principle of preserving the significant additions of later periods, demanded by Articles 2 and 11 of the Charter. Article 8 provides that "the sculptural, painted or decorative elements, which are an integral part, inseparably bound to the monument, may not be separated from it, unless that is the only way of insuring their preservation." (Fig. 25). Thus the sculptural decoration of the Parthenon, the Erechtheion, and the temple of Athena Nike were removed from the monuments when research on the mechanism of marble erosion showed that this was the only escape route for assuring their preservation. According to Articles 9, 12 and 15 the fillings of new marble are restricted to the absolutely necessary, precisely correct morphologically and distinguishable from the authentic parts (Bouras 1994, 2007).

Figure 22.   Removal of the iron beams incorporated into the marble beams during the previous restoration.

Figure 23.   Joining the fragments of a beam by means of titanium rods.

Figure 24. Parthenon. Resetting of an architrave.

Figure 25. The copies of the removed from the Parthenon Opisthonaos west frieze blocks after placing them on the monument.

Especially for the monuments of ancient Greek architecture and in the framework of studies for the restoration of the Acropolis monuments, additional principles were formulated especially for the monuments of this time.

- The principle of reversibility of the interventions, that is, the possibility of returning the monument to the condition it was in prior to the intervention, so that all possibilities for information are preserved and so that any error in today's interventions can be corrected in a future intervention.
- Basic criterion of the interventions for the structural restoration is the respect for the authentic material, retention of structural autonomy of the architectural members and their original structural function.

These general principles lead to specific scientific choices, technical solutions and construction details which formulate the frame of the specific principles and procedures in the designing and application of structural interventions. These are:

- *Restriction of interventions to the absolutely necessary.* The interventions are carried out on those parts of the monument where failures are apparent that could lead to the danger of collapse, breakage or cracking of members or large shifts with the danger of yielding. The undisturbed areas of the monuments are left intact so as to preserve fully their authentic character and to provide future generations with the wealth of information that they hold.
- *Respect for the original structural system of the monuments.* Since the excellent seismic behavior of the monuments is due to their original structural system, retention of that system in all its detail is of utmost importance. If in the course of the interventions the resistance and stability of the bearing structure are altered, the behavior of the areas of intervention is affected in case of seismic strain. The result would be undesirable for the entire structure. A characteristic example is the structural restoration of the columns in situ in the Parthenon Opisthonaos, which had been cracked by thermal fracture during a fire in ancient times. It was decided to fill the cracks with suitable hydraulic grout. This solution was chosen as opposed to dismantling, in order to avoid disturbing the original structure of the columns that was still preserved. In order to maintain the structural integrity of each drum after grouting, since the grout would necessarily pass through the contact surfaces of the drum, special research was carried out with the following results: two different grout compositions were applied, one within the drum developing great resistance and bonding force with the ancient marble and another in the vicinity of the joints between the drums with very weak initial resistance and bonding force with the marble, weakening further in time, in order to avoid bonding between drums in the contact area.
- *Preservation of the original structural function of the architectural members during their restoration* (Ioannidou et al., 2003). Broken architectural members are restored to their original monolithic form by bonding the joining fragments with titanium rods and white cement. Because the members are

28

placed on the monument in an isostatic system, each member can be considered as an integral unit in relation to the entire construction. Wall blocks are subject to single-axis compression, column capitals to bending and shearing and beams and architraves mainly to bending. The principle of respect for the original structural function and the autonomy of the architectural members is applied so that the strength of the mended member is at most to that of the unharmed member. Thus, in joining fragments of members, we attempt to use as few as possible of the titanium reinforcements demanded by the structural study. Likewise the holes (sockets) for the rods are as small as possible so as to avoid damage to the ancient material. Members that have been subjected to bending can be structurally restored to their original resistance. If in this case the calculations result in very heavy reinforcements, thus considerable loss of ancient material, the resistance of these members is restored to withstand the total strain that may apply after the anastelosis of the monument. These forces include the structural vertical loads and the powerful seismic vertical and horizontal shocks. Critical for calculating the bonding are the stresses developed on the broken surface. Analysis of the stress-condition of the combined structure of marble-titanium is performed by methods used in classic structural calculations, whereas distribution of the strain is determined by applying the general principles of the strength of materials. For determining the strength of the bonding, the strains developing in the bonded crack are compared with the allowable strains of the material, i.e. the pressure of the marble and the tensile strength of the titanium. The amount of reinforcement determined by this method is very small (1–2%) and damage to the ancient material is thus slight. Particular attention is given to the width of the joint in the area of bonding so that it is as small as possible and therefore less mortar is used. To obtain this result, the bonded fragments are kept well-tightened during bonding and during the first days of curing of the mortar.

- *Filling in the ancient members with new marble.* The required restorations in new marble are usually limited and the criterion for the decision is always the stability of the member and the monument and the structural and aesthetic self-autonomy of the areas being restored. The surface to be joined is cut, using a pointing device or a copying machine so that the new piece can match precisely with the broken surface of the ancient piece (Fig. 26). This method complies with the principle of reversibility. Thus, if the missing ancient piece is identified in the future, it will be possible to remove the modern filling and re-set the original fragment in its place. For the actual bonding of the filling of new marble to the ancient piece, the same procedure applies as described above for the joining of ancient fragments.

- *Joining the blocks.* The restoration of the joining elements that have been damaged by stress or by the rusting of the iron is an intervention that contributes to the resistance of the building in seismic strain. In case of earthquake, it is the clamps and dowels that assume the load and, together with the friction bond, they reduce the relative displacement of the marble blocks. The new clamps and dowels are made of titanium sheet and they are secured in the ancient sockets with cement mortar (Fig. 27). These metal reinforcements are designed to be the weaker element so that in cases of great strain they can yield to permanent deformation and, if a new intervention proves necessary, it can be limited to replacing the clamps/dowels. Wherever the ancient cuttings are preserved, I—shaped clamps are set, following the form of the ancient clamps with heads that are thicker in relation to the stem. The long stem has a neck so that when the clamp is under tensile stress and the material exceeds its limit of yielding, it will be deformed and will therefore absorb a considerable amount of energy before it fails. The space between clamp and cutting and within an area of 1–2 cm on both sides of the joint is left free, without cement mortar, to allow room for possible crosswise deformation of the clamp. If the original socket has been destroyed, two blind holes are drilled at the position of the original heads of the I—clamp. The holes are made to receive the vertical legs of the new clamp, which is now Π—shaped. If the area where the clamps should be attached is broken, specially shaped joining clamps are designed. The designing of dowels is handled in similar fashion.

- *Restoration of parts of the monuments by resetting the ancient members in their original positions and including also members of new marble.* Apart from considerations of general principles of theoretical conservation ethic, these restorations are made on the basis of structural criteria. An increase, for example, of vertical load and the accompanying increase in friction between the stones, the

Figure 26. A new marble filling to a column capital.

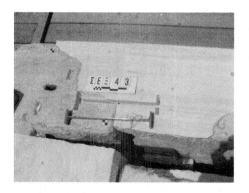

Figure 27. Setting a double—T clamp of titanium for joining two marble blocks.

supplementation of the original design and regularity of the ground plan of the monument, an increase of rigidity with the restoration of the walls and supplementation of the ceilings and roof, which add to a partial diaphragmatic function of the solid disk of the ceiling, are all favorable in assisting the building to withstand seismic forces. Relatively characteristic examples are seen in the restoration of the coffered ceilings of the central building of the Propylaia (Fig. 4).

- *Quality of construction.* It is particularly important that a high quality of construction be maintained in the interventions in all phases of the work, for both structural and anti-seismic reasons. As in the ancient construction, exceedingly careful working of the contact surfaces and the perfect contact between the blocks insures the development of friction forces between them and through them the cohesion of the building. This concern, quite apart from aesthetic demands, is necessary for purely structural reasons, namely respect for the ancient structural system of the monument, which thus insured the safe transfer of the load to the ground.

4 SEISMOLOGICAL ASPECTS

4.1 *Seismotectonic regime of the area*

The seismicity of the broader Aegean area is caused by the relative motions of the Africa, Arabia and Eurasia plates. The area is not rigid and hence is not part of any plate (Ambraseys, 2010). Greece and the Aegean sea is one of the most rapidly deforming continental areas, therefore they have been much studied. The Attiki peninsula region, in central Greece, is considered as an area of low seismicity. The tectonic regime of Attiki is better understood if it is examined together with that of the adjacent areas to the west (Eastern Corinth Gulf) and to the north (Thiva basin, Evoikos Gulf, Oropos area) where strong earthquakes occurred in the past. These areas are associated to an extensional N-S tectonic field and the major faults, which can produce a significant earthquake end gradually before they reach the Attiki district and furthermore Athens area. The tectonic field and the seismicity is described by various authors (e.g. Galanopoulos 1971, Ambraseys & Jackson 1997, Papazachos & Papazachou 2003, Rondoyianni et al., 2000). A comprehensive review on the long-term seismicity of Athens was presented by Ambraseys (2010) in Academy of Athens. Near the historical center of Athens, within a radius of 10 km only small, shallow normal faults exist, which are not capable to produce any earthquake. All the known active faults are more than 10 km away from the center of Athens and are of short length, implying that the earthquakes originated by them are only felt in Athens, with no damage at all. So, it seems from the systematic historical studies and the recent earthquakes that the seismic hazard in the center of Athens is rather low. However, Ganas et al., (2005) concluded that events with M_s between 6.4 and 6.6 are to be expected from possible movements of the Attiki mapped faults.

Athens suffered from historical and instrumental earthquakes from these areas. A short description of these earthquakes and their effects to Athens and specifically to Acropolis are mentioned here (Galanopoulos 1955, Papazachos & Papazachou 2003, Ambraseys 2010):

427-426 BC, Orchomenos (Imax VII). Thucydides mentioned that during that period strong earthquakes occurred in central Greece. Although Korres & Bouras (1983) mentioned that due to these earthquakes Parthenon suffered some damage, the epicentral distance of 100 km from Athens and the lack of historical information for severe damage in the city of Athens leave a doubt for the real cause of this.

1705, Oropos (Imax VII). Sieberg (1932) and Galanopoulos (1955) mentioned that this earthquake occurred in 1641. Ambraseys (1996) based on Turkish documents places the earthquake on September 3rd, 1705 and Papazachos & Papazachou adopt this opinion. The epicenter is placed to the north of Athens at Oropos to a distance of about 40 km. Many churches and official buildings around the center of Athens suffered severe damage. According to Korres (1996) this earthquake caused the collapse of the east retaining wall of Acropolis. Also, the majority of the Acropolis cisterns were ruined.

1785, Oropos (Imax VIII). Although serious damage occurred at Oropos and Chalkida, very little information exist for this earthquake, which was only felt in Athens.

1805, Athens (Imax VII). The earthquake caused damage at Parthenon. It has been suggested that this earthquake (on the one of 1785) caused the collapse of a part of the north Circuit Wall (Egglezos and Moullou 2011).

1874, Athens (Imax VII). Galanopoulos (1955) mentioned that this earthquake (January 17th) caused damage in Athens and at Acropolis. Probably it is the same earthquake that Papazachos & Papazachou (2003) and Ambraseys (2010) refer to it as of March 18th.

1894, April 27, Atalanti (Imax V). The epicenter was located at a distance of 80 km to the north of Athens. Important damage occurred at the meizoseismal area due to a strong foreshock. Historical churches and monasteries were destroyed. Fissures were observed at some houses and buildings in Athens. Parthenon pediment suffered damages.

1938, July 20, M6.0, Oropos (Imax VIII). The earthquake (at a distance of about 40 km to the north of Athens) caused serious damage at the epicentral area villages. 18 persons were killed and landslides occurred at various sites.

1981, February 24, M6.7, East Corinth Gulf (Imax IX). This strong earthquake caused the damage of 500 houses in Athens, mainly old houses and multi storey buildings. Some small slides and joint widening occurred at Parthenon. The surface trace of the seismogenic fault reached the length of 15 km and an average drop of 60 cm (Papazachos & Papazachou 2003).

1999, September 7, M5.9, Athens (Imax IX) The most recent earthquake caused 143 deaths, about 800 injuries and about 40 collapsed buildings (Elenas, 2003). The seismogenic fault has not a surficial appearance, and although the epicenter was about 20 km from the center of Athens no serious damage occurred at Acropolis.

Table 1 summarizes the main historical and instrumental earthquakes that affected the Acropolis and the monuments taking into account the literature search (Galanopoulos 1960, Papazachos & Papazachou 2003) and using attenuation relations (Skarlatoudis et al., 2003) to estimate PGA. Figure 28 shows a geological map of the hill (Higgins & Higgins 1996).

Table 1. Earthquakes that affected Acropolis.

Year	Site	Distance (km)	M	PGA* (cm/s^2)	Effect
426 BC	Atalanti	140	?	?	Parthenon
1705	Oropos	35	> = 6.0	45	South Wall
1785	Oropos	40	> = 6.0	40	North Wall**
1805	Oropos	40	?	40	Parthenon, North Wall**
1874	Athens (?)	?	?	60(?)	North Wall
1894	Atalanti	100	6.9	30	Parthenon
1938	Oropos	37	6.1	40	South Wall
1981	Korinthos	77	6.7	30	Parthenon
1999	Athens	20	5.9	65	Propylaia

*estimated, ** Egglezos and Moullou, 2011.

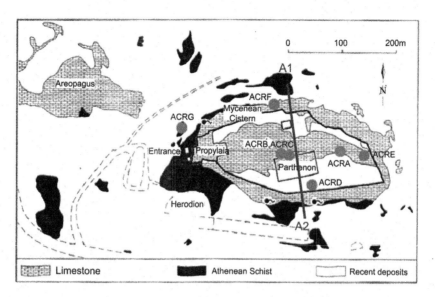

Figure 28. Geological map of the Acropolis hill (revised from Higgins & Higgins 1996), showing the main topology names, the accelerograph installation sites (grey circles) and the section A1 A2 of Figure 31.

It should be noted that the selection of the proper seismic action for the design of any intervention on monuments, besides the obligations arising from the current regulations and codes (e.g. EC8), should also take into account historical seismicity data. This procedure allows for a rational approach meeting the needs for safety as well as for the implementation of restoration principles.

4.2 Macroseismic observation

The occurrence of intermediate depth earthquakes not only within the broader Aegean area but close to Athens as well should be mentioned. In 1964, July 17th a strong earthquake (Ms = 6.0) occurred near Athens, about 15 km to the west of the city center. Due to its intermediate depth (150 km), the main effects appeared at Elia and Messinia prefectures of West Peloponnese, at a distance of more than 150 km (Fig. 29, left). On the other hand, the near-field strong earthquakes of shallow focal depth affect the urban area of Athens, as that of September 7th, 1999 with about the same magnitude (Fig. 29, right).

4.3 Accelerographic array

The deployment of Acropolis strong motion array started in 2006, after the Acropolis Monument Restoration Service (YSMA) expressed interest. NOAIG established a digital 3-component accelerograph on the Acropolis hill (emergence of limestone) with a 12 bit resolution (ACRA). The array was finalized after 2008, when Institute of Geodynamics of the National Observatory of Athens (NOAIG) installed six digital 3-component accelerographs of the latest technology with a 24 bit resolution. Six sites were chosen for installation in order to cover various local characteristics: a site (ACRG) on the Athenian schist (geological bedrock), a site (ACRF) on the limestone at the base of the North Wall, a site (ACRE) on the east side of the hill (on limestone next to the Wall), a site (ACRD) on the south side of the hill (where there is the thickest artificial fill), a site (ACRB) at the base of one of the columns of the Parthenon north colonnade and a position (ACRC) on the architrave of the same Parthenon column. Figure 28 shows the installation sites of the acclerographs.

Details concerning the installation techniques as well as different earthquakes recorded by the array are described in Kalogeras et al., (2010), Kalogeras et al., (2012) and Kalogeras and Egglezos (2013).

The continuous streaming of the modern Acropolis accelerographic array provides the ability to monitor the long-term seismic noise at each site, by using the PASSCAL Quick Look eXtra (PQLX)

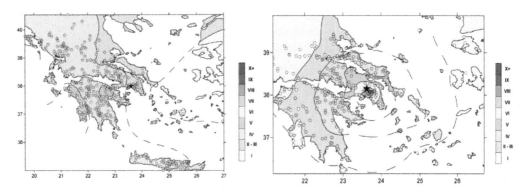

Figure 29.   Comparison of the macroseismic field of two earthquakes occurred near the Athens center having the same order of magnitude but with different focal depths. At the top the higher intensities occurred at a distance more than 150 km (intermediate focal depth), while at the bottom the higher intensities appeared near the epicenter (shallow focal depth). Maps from Schenkova et al., 2003.

software tool (McNamara & Boaz 2011) based on the McNamara & Buland (2004) approach. Useful conclusions extracted showing the variations due to the different local geological conditions at each installation site, as well as the differences between ACRB and ACRC ambient noise records reflecting the Parthenon response (Kalogeras et al., 2012, Kalogeras and Egglezos 2013).

Although no near-field moderate or strong earthquake occurred during the operation of the Acropolis accelerographic array, the earthquakes recorded can be classified into three main categories: a) weak local earthquakes with shallow depths, having a local magnitude Ml between 3.0 and 3.5 and epicentral distances no more than 50 km, b) moderate regional earthquakes with shallow depths having an Mw between 5.0 and 5.5 and epicentral distances of about 250 km and c) moderate regional earthquakes with intermediate depths (~ 60–130 km) having magnitudes Mw of the order of 6.0 and epicentral distances of about 350 km. The general conclusions resulted from the processing of these records showing the effect of local installation conditions for the local earthquakes (the higher ground acceleration was recorded at the artificial earth-filling—ACRD—site), while the site amplification does not play a significant role for the moderate or strong regional earthquakes, independently of their focal depth (Kalogeras and Egglezos 2013).

It is worth to mention the characteristic dominant response period of 0.5 s, which is apparent in the response spectra of all the records of ACRC site. Figure 30 compares the response spectra for the ACRC, N-S component for 3 representative earthquakes of the fore-mentioned categories (shallow depth, local to the left; shallow depth, regional in the middle; intermediate depth, regional to the right). Red triangles show the tail part of the record, which was processed. This specific period of 0.5 s is also apparent at the response spectra of ACRC whole records, especially for the intermediate and shallow depth regional earthquakes, while for the shallow depth local earthquakes it is not the dominant period (Kalogeras and Egglezos 2013).

4.4   *2-D Dynamic response analysis*

The execution of preliminary 2-D dynamic response analyses of the broader area subsoil model of the Acropolis rockhill aims to reliably predict the seismic motion along the free surface of the sacred rock—terrace and slopes (Kalogeras and Egglezos 2013). Within this context, the analytical results were compared with the recordings of the Acropolis accelerographic array for different seismic scenarios, aiming at calibrating the dynamic properties of the geomaterials encountered in the rockhill and at evaluating the modeling applied for the analyses.

The analyses were conducted along the middle cross-section of the rockhill, in the N-S direction (Figs 28 and 31), the layering of which was defined according to the existing geological data (Andronopoulos and Koukis 1976). The dynamic properties of the geomaterials were assigned according to the relevant

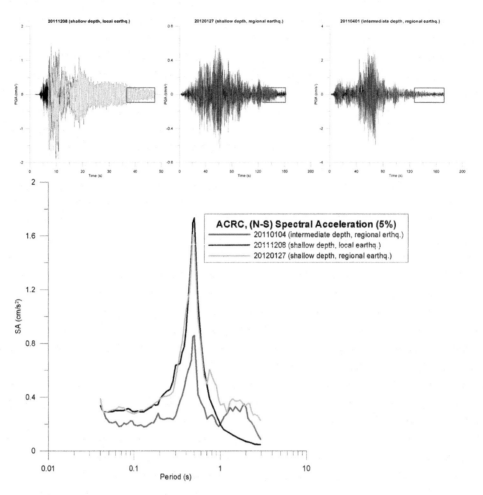

Figure 30. Comparison of the spectral acceleration for the N-S component of ACRC (Parthenon architrave) for the tail part of representative records. A dominant frequency of 0.5 s is clearly apparent independently of the earthquake characteristics (magnitude, epicentral distance, focal depth).

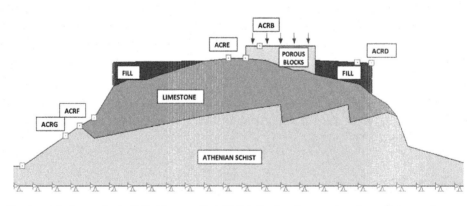

Figure 31. Characteristic section of the Acropolis hill (N-S), for the conduction of dynamic response analyses.

literature (NEHRP, 2005) for geological formations of similar nature and are presented in Table 2. The modelling applied for the dynamic response analyses, though its simplicity, seems promising for the prediction of the superficial seismic motion in the Acropolis rock-hill. However, a more detailed modelling of the in situ encountered geological zones and geomaterials and their dynamic properties (e.g. from the results of a properly focused geotechnical survey) could be exploited for a systematic parametric set of 2-D analyses and significantly improve the accuracy of the results.

34

Table 2. Dynamic properties of geomaterials.

Material	Vs (m/s)	Gmax (kPa)	v	ξ (%)	γ (kN/m³)
Limestone	1500	6.08E+06	0.25	10	26
Schist	700	1.20E+06	0.25	5	25
Backfill	350	2.45E+05	0.30	1,5*	20
Poros	600	8.64E+05	0.30	40**	24

elastic property (ξ_{min} (%)) ** notional value to account for the Parthenon foundation structure.

5 GEOSTRUCTURAL STUDIES AND INTERVENTIONS ON THE ACROPOLIS

5.1 Available geotechnical data

Systematic geotechnical research has never been conducted on the Acropolis plateau. Although, in the past years there have been several suggestions for the necessity of such an investigation, the Acropolis Committee did not authorize it. The main argument is that the borehole drilling of the artificial fill could permanently damage ancient architectural or sculptural fragments that may remain buried in the backfilling of the plateau.

Therefore all the data used in the design of geotechnical interventions on the Acropolis summit come from limited available laboratory data, assumptions based on general literature and macroscopic observations evaluated through back analysis, non destructive in situ tests and monitoring recordings. This difficulty in attaining raw data leads, inevitably, to a conservative design approach.

Nevertheless, the recent damages (Feb. 2013) in areas of the façade of the south Circuit Wall and the rockfalls of the north slope, due to heavy rainfall, opened again the discussion for the implementation of a small scale investigation project as a prerequisite for the design and application of stabilizing measures. The realization of such an investigation will enrich our knowledge for the geomaterials encountered in the Acropolis area.

5.1.1 Data from laboratory tests
In this paragraph the available laboratory geotechnical data are presented. The existing data concern the properties of the natural geomaterials encountered in the broader Acropolis area and the stone blocks of structures with a prevailing geotechnical nature (the foundations of the monuments and the retaining Circuit Wall).

These data origin from the following sources:

- The geological study of the Acropolis (Andronopoulos and Koukis, 1976). In this study limited laboratory tests on the rocky materials are included (limestone, schist and conglomerate formations).
- The study for the conservation of the Arrephorion foundation walls (Maridaki et al., 2006), concerning the marly and dolomite stone structural blocks.
- The laboratory tests on poros stones fragments in terms of a project concerning the joining of the poros fragments (Theoulakis et al., 2009).
- The exploratory drillings during the project of stabilization of rock pieces in the SE area of the limestone rockhill (December 2008). During this project rock samples of superficial conglomerate were tested.

Concisely the above mentioned data are summarized in Table 3.

5.1.2 Other sources of data of geotechnical interest
Besides the pure quantitative data from laboratory testing it is of value to mention characteristic qualitative data that may be used to supplement the existing geotechnical information.

Table 3. Available data from laboratory tests. Synopsis.

ORIGIN	Site	Description	Compressive strength q_u (Mpa)	Point load test Is50 (Mpa)	Brazilian test σ_t (Mpa)	Triaxial Compression c(Mpa)	φ(°)	Flexural strength—three points (Mpa)	Elastic Modulus E (Gpa)	Poisson ratio v	Density (gr/cm³)	Hardness Mohs
Andronopoulos and Koukis, 1976	Bedrock	marly schist	33.5	–	–	–	–	–	–	–	2.51	4
		fine grained marly Sandstone	11.6	–	–	–	–	–	15	0.21	2.54	4
		"	38.2	–	–	–	–	–	62.5	0.16	2.58	4
		coarse grained marly Sandstone	31.5	–	–	–	–	–	25	0.15	2.61	4
		"	24.6	–	–	–	–	–	25	0.17	2.62	4
		"	12.3	–	–	–	–	–	25	0.22	2.57	4
	Rockhill (Slope)	Limestone	40.1	–	–	–	–	–	10	0.29	2.67	4
		Conglomerate	45.5	–	–	–	–	–	27.5	0.29	2.67	4
		"	23.3	–	–	–	–	–	62.5	–	2.48	4
		"	25.3	–	–	–	–	–	–	–	2.59	–
Stabilization of small rockpieces in the S-E corner, 2008	Rockhill S-E slope	Conglomerate	24.65	2.54	–	–	–	–	1.87	0.21	2.5	4
		"	27.09	2.61	5.79	–	–	–	2.11	0.16	2.48	4
		"	–	–	5.12	–	–	–	–	–	2.48	4
		"	–	–	–	6.88	33.3	–	–	–	–	–
Project for joining of poros fragments, 2009	Erechtheion Dolomite		42.74	–	–	–	–	–	–	–	–	–
	"		42.6	–	–	–	–	–	–	–	–	–
	Scattered	"	–	–	–	–	–	7.11	–	–	–	–
	"	"	59.28	–	–	–	–	9.91	–	–	–	–
	General literature	"	68.5	–	–	–	–	9.45	–	–	–	–
	"	"	95–113	–	–	–	–	–	–	–	–	–
	Erechtheion Marl	"	7.59	–	–	–	–	11.7	–	–	–	–
	Scattered	"	6.57	–	–	–	–	14.32	–	–	–	–
	"	"	20.47	–	–	–	–	10.29	–	–	–	–
	"	"	18.93	–	–	–	–	13.43	–	–	–	–
	"	"	–	–	–	–	–	1.45	–	–	–	–
	"	"	11.35	–	–	–	–	–	–	–	–	–
	"	"	5.92	–	–	–	–	–	–	–	–	–
	General literature	"	11.5–24.5	–	–	–	–	–	–	–	–	–

- Qualitative description of geomaterials from excavation data from the relevant archaeological reports (e.g. Cavvadias and Kawerau, 1907).
- Photogrammetric recording and 3-D laser scanning of the Circuit Wall and the rockhill, providing the necessary data a) for the accurate modeling of the hill and the Wall needed for the conduction of a reliable 2-D or 3-D analysis, b) for the identification of the Wall's structural pathology (Moullou and Mavromati, 2007).
- Geophysical survey data of characteristic sections of the South Wall and the retained backfill (Tsokas et al., 2006). This project aimed originally at determining the Wall's geometry at predefined sections. Although the main goal was not achieved, however, interesting information relating to the water regime behind the Wall emerged. This information allowed for a more realistic estimation of the parameters impacting on the Wall's structure: saturation of soil backfill, increase of earth pressures, smoothing of the joints between stone blocks (decrease of frictional forces), potential development of hydrostatic pressure behind the Wall (especially in cases with an external revetment from masonry of low permeability), loss of mass due to the erosive action of the penetrating water. The concentration of water content in characteristic sections of the South Wall is depicted in Figure 32 a,b.

5.1.3 *Data obtained through analytical applications and monitoring recordings*

In this paragraph sets of geotechnical parameters (values) that were used in various geotechnical analyses on the Acropolis monuments are presented. Specifically, the aforementioned data were validated a) through back analysis for the interpretation of a permanent structural failure in the north Circuit Wall (Egglezos and Moullou 2011), b) through 3-D analysis for the assessment of the Parthenon foundation response with exploitation of in situ tests (horizontal loading) on the columns of Parthenon's Opisthodomos (Egglezos and Toumbakari 2011) and c) through analyses for the interpretation of systematic monitoring recordings (Egglezos 2010a).

5.1.3.1 *Back-analysis of existing damages and failures for the assignment and/or calibration of mechanical parameters values*

This work concerns a back-analysis for the interpretation of a serious structural failure observed in an area of the Acropolis north Circuit Wall (Egglezos and Moullou 2011). The failure occurred, according to the available historical and archaeological evidence, at the end of the 18th or beginning of the 19th century and it includes a) the collapse of the upper part of the Wall in the area examined (the crown of the Wall that had been constructed of architectural members of the entablature of the Old Temple of Athena), b) significant outward lean from the vertical (7 cm) of the remaining lower part (the part beneath the crown that collapsed), c) rotation of approximately 1°, d) systematic cracking of the outward face of the Wall. For the analysis, the principles of rock mechanics were applied for the simulation of the dry-masonry construction. Since the damage was primarily attributed to a strong earthquake motion, for the problem description information was drawn from a wide range of scientific specialization. The analyses comprised: 2-D FE elastoplastic analysis, simplified (pseudostatic) spectral

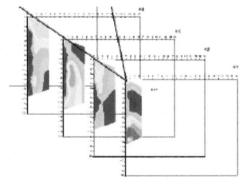

Figure 32.   Geophysical survey data (a) showing the concentration of water content in characteristic sections of the South Wall (b).

dynamic analysis and a rigorous 2-D time history analysis. The results are proven very satisfactory. The permanent displacement of the current crown of the Wall, the rotation of the structure, the collapse of the entablature members (form Γ) and the systematic net of fractures at the outward side of the Wall are consistently predicted (Figs. 33–34). Therefore, the assumption that the damage was caused by an earthquake was also verified, limiting the termini of the event between 1785 and 1805.

The geomaterials involved in the analysis are the following:

- Limestone bedrock: the unit upon which the Wall is founded. Rock mechanics properties are defined according the macroscopically estimated GSI (Hoek et al., 2002, 2006, V. Marinos et al., 2005) and uniaxial compression values from tests on intact rock (Andronopoulos and Koukis 1976).
- Backfill soil material. According to the existing archaeological data (Cavvadias and Kawerau 1907) and to the observations made on the spot by the authors, this artificial material could be described on geotechnical terms, as a well compacted clayey Gravel (GC).
- Structural stone blocks (Marl to marly Limestone). They come from the geological rocky unit met at Piraeus coast (known as Aktite). The definition of their mechanical properties derive from existing data from uniaxial compression tests ($7.01 < q_u < 24.4$ MPa) (Theoulakis et al., 2009) and the macroscopically estimated GSI. The application of GSI accounts for the preservation state of the stone blocks.
- Contact surfaces of blockstones. The geotechnical parameters of contact surfaces (open joints of marl, with slight roughness) are obtained (due to the lack of data) from the relevant literature (Wyllie 1999).

The mechanical parameters of geomaterials are shown on Table 4.

Table 4. Geotechnical values used for the back-analysis.

	GSI	c	φ (°)	E_{RM} (MPa)	q_u (MPa)	v	γ (kN/m³)
Marl	50*	100	40	460	12	0.2	24
Limestone	50–70	–	–	7000	40	0.2	26
Backfill	–	5	40	100	–	0.3	20
Contact surface	–	0	25	–	–	–	–

*To account for ageing effects (current preservation state of the stones) of the structural block strength.

Figure 33. North Wall. Detail.

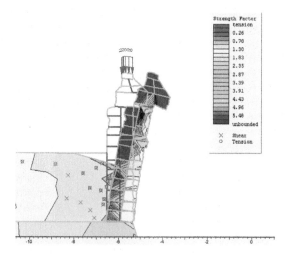

Figure 34. Interpretation of the damages in the wall face (Shear strength ratio).

5.1.3.2 *Utilization of in-situ full-scale experimental measurements upon the Parthenon's Opisthodomos columns*

This project concerned a) the in-situ trial load tests, which were carried out on the four, out of six, columns of the Parthenon western side, the Opisthodomos, during the recent restoration project (2001–2004) and b) the analytical assessment of their structural response (including foundation) based on the obtained experimental data (Egglezos and Toumbakari 2011).

The dismantling of the entablature members during the anastylosis project, gave the opportunity for the performance of in situ trial load tests in order to study the response of the columns in interaction to their foundation. The testing project consisted of the application of a horizontal force of 10–11 kN at the capital of each column. During the test, the horizontal displacement of the capital and the vertical displacement of the base marble block, were measured. The test was carried out twice in each direction, to account for differences in the foundation conditions. For the evaluation and the interpretation of the test results a back-analysis was performed. For the (3-D nonlinear elastic) analysis Rock-Mechanics principles were applied. The idea was that dry masonry structures from natural rock stone blocks connected with frictional "joint" forces are analogous to a jointed rockmass system. The comparison of measured values and analytical predictions were generally in good agreement, offering increased accuracy. According to the results, discrete modelling with properly determined parameters for monument's geomaterials can satisfactorily be applied to restoration analyses of this type of structure and could form the base for more complex calculations involving dynamic effects e.g. earthquake.

The geomaterials involved for analytical interpretation of the results are the following:

- Marble drums and Crepis
- Poros Footing Blocks (marly limestone)
- Dolomite Footing Blocks
- Limestone Bedrock.

The footing blocks came from the rocky geological formation of the Piraeus coast, known as Aktite, which appears typically either as a soft rock (marl to marly limestone) or as a dolomite.

The assignment of mechanical and geostructural properties to the above materials (Young Modulus E, Poisson ratio v) is a prerequisite for the conduction of the analytical calculations. Since these materials are classified as rock, the application of GSI method (Hoek et al., 2002, 2006, V. Marinos et al., 2005) applies for the estimation of the rockmass properties with the use of Roclab software code (Rockscience Inc. 2005), according to the methodology mentioned in the previous paragraph. The initial mechanical parameters for application of the GSI method (uniaxial compressive strength q_u and E of intact rock), are obtained from relevant literature (Andronopoulos and Koukis 1976, Theoulakis et al., 2009, Egglezos and Moullou 2011), while the value of GSI is based on expertise macroscopic estimation. The initial parameters and the results for rockmass properties are summarized in Table 5.

5.1.3.3 *Data arising from monitoring recordings*

Additional sources for the production of useful mechanical parameters are the recorded values from the installed monitoring systems for the observation of the Circuit Wall performance.

Specifically, characteristic elastic properties of the Circuit Wall were extracted from the utilization of topographical and optical fiber sensors measurements for the monitoring of the south Circuit Wall response.

Table 5. Mechanical properties of Geomaterials.

	Intact rock				Rockmass	
	E (MPa)	v	q_u (MPa)	γ (kN/m³)	GSI	E_{RM} (MPa)
Marble	70000	0.20	80	27.0	70	50000
Marl	2000	0.25	11	25.0	60	1000
Dolomite	18000	0.20	60	26.0	60	13000
Limestone	50000	0.20	100	26.0	70	35000

The measurements were used for validating the 3D model of the SE area of the Wall (made of marly stone blocks). The validation concerned the comparison of measurements and analytical results of the strain field in the south area of the Wall, arising from the action of the temperature load that corresponds to summer heat. For the 3-D analysis the following values were assigned to the applied parameters:

E = 500 MPa, a = 5*10–6, v = 0.20

The model, simplified on purpose, contains the following:

- Elastic analysis (choice supported by the systematically alternating sign of the measurements).
- Idealized geometry of the Wall.
- Documentation of the Wall as a continuum with reliable properties of the masonry body.
- Sufficient length so as to limit, to the extent possible, the effect of the bordering restraints.
- Temperature load (because of change in the environmental temperature) of 20° C, given that the reference measurement was taken in November.

The strains resulting from the analysis (Fig. 35a) were in clear agreement, from the standpoint of both qualitative (lifting, transverse movement, longitudinal movement from E to W) and quantitative level with the data derived from the topographical measurements of the same section of the Wall (Egglezos 2010a). Thus the modeling used for preliminary analyses was considered sufficient.

In the context of further validation of the model, the results from the optical fiber sensors that had been placed on the upper part of the South Wall, in E-W direction were also used. The calculations for the strain and stress field (Fig. 35b) of the Wall area agree with the results from the measurements.

The above evidence enforces the estimation that the model of the SE area of the Wall (concerning the stresses to which it has been subjected, and the properties of the materials), is good and can be used as a base for more complex analyses.

5.2 Geotechnical studies and interventions

In the following paragraphs the main interventions with prevailing geotechnical nature are presented. These interventions refer to the stabilization of the rockhill slopes, the rescue interventions for the preservations of the existing ancient monuments and the construction of modern infrastructure necessary for the implementation of the restoration project.

5.2.1 Rock slope stabilization interventions

Due to continuous rockfalls from the Acropolis rockhill emerged the need for systematic implementation of stabilizing measures. The measures that have been applied during the current restoration project, concern either the stabilization of bulky rock blocks or the consolidation of the highly weathered rock surface in the southern and south-east part of the rockhill. Large parts of the North Slope

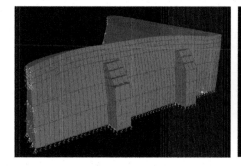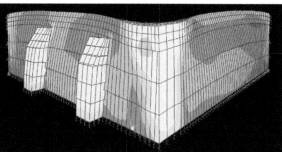

Figure 35 a,b.   Results of 3-D analysis. Strain field (a) and stress field (b) of the Wall, due to thermal loading.

remain unsupported. It is worth mentioning that in the north Acropolis slope the limestone is highly weathered and tectonized (compared to the other slopes of the hill). It is clear that for safety reasons, concerning the visitors and employees, as well as the stability of (parts of) the Circuit Wall, the rock stabilization project should be applied to all the remaining unconsolidated areas of the hill.

5.2.1.1 *Stabilization of the rock blocks in the S-E area of the rockhill*

During the period between 1977 and 1993 extended stabilization measures were applied for the consolidation of the Acropolis rock slopes (Monokrousos 1995). The stabilization concerned either unstable bulky rocky blocks or weathered and fractured areas of the slope. The fracturing of the rockmass is attributed to the tectonic regime of the site, the disorganizing action of the plant and trees roots and the weathering from the continuous action of natural agents.

The stabilization project comprised the classification of the slope areas in classes (according to the rockfall hazard) and the implementation of stabilizing measures. The typical measures applied are listed as following:

- Implementation of active prestressed anchors of various lengths (2–4 m for small rock pieces, 6–16 m for medium rock blocks, and 20–24 for huge rock blocks)
- Application of grouting for the filling of open joints
- Cutting of the plants so as the unfavorable wedge action of the roots is ceased
- Construction of stone columns for the support of hanging rock masses.

The project has been applied to almost the 2/3 of the Acropolis slope surface (covering particularly the south, the east, and the northeast side of the hill), thus offering the required safety for the visitors and the stability for the monuments that are influenced from the rock slope instabilities (mainly the south and the east Circuit Wall).

5.2.1.2 *Stabilization of small rock pieces in the S-E corner of the rockhill*

The project concerns the Southeast slope of the Acropolis Rockhill. This site is located at the border of intensely tectonized limestone and the appearance of conglomerate formation. The rock mass is highly and densely jointed, thus creating low volume rock blocks. The discontinuities of the rockmass are generally open with a fill of clay material. The morphology of the area creates an unfavorable situation since this intensely weathered area belongs to an overhanging part of the rock relief. Although the general stability of the slope is secured due to the extended stabilization measures applied in the past (see the previous paragraph), the superficial zone of the slope (30–50 cm thick) contains numerous instable (in limited equilibrium) rather small size rock blocks. Indeed, the falling of rock pieces is a typical consequence after high rainfalls, leading to further loosening of the rockmass.

For the treatment of this evident risk an intervention for the stabilization of the area took place (2008), comprising the following measures:

- Deep cleaning of the rockmass joints and grouting/mortar application (by hand) for the creation of a continuous mass.
- Grouting application through drilling holes for the reinforcement of the outer layer of the rock, in a depth of about 1.5 m.
- Installation of metallic bolts (in the above mentioned holes) in an appropriate pattern for the stabilization of the rock mass and the suspension of the instable superficial zone to the deeper sound rock.

The application took place in an area of about 120 m². The works in the slope were performed through a wooden scaffolding aside the rock slope.

The bolts were applied in a density of 1 piece/1.7 m², with working load capacity of 18 kN. The dimensioning of the bolts took into account the Von Mises failure criterion. The calculations were performed for bolts 60 cm long with assumed fixed support equal to 20 cm, for steel AISI316 (f_{yk} = 240 MPa), for nominal hole diameter equal to 20 mm and minimum adhesion between the supporting element and the surrounding rock equal to τ_r = 3 MPa. From the design calculations a bolt diameter Φ = 14 mm resulted.

The holes for the bolts (before their introduction) were used for the low pressure (0.30bar) grouting of the fractured rockmass, aiming at a (nominal) filling of discontinuities in a depth of about 1.5 m.

The mortar for the superficial sealing was prepared from white cement and limestone oriented sand in a volume ratio C/S = 1. Proper dyes were added for a successful adaptation of the intervention in the natural environment. The water quantity was determined so as the mortar remains in plastic state. The grouting was prepared from white cement in a (weight) ratio W/C = 0.6.

The geotechnical conditions for the application of the stabilizing dowels were investigated through two parahorizontal (inclined) sampling boreholes (4 m long each). These boreholes served also as draining holes of the sealed (due to the grouting) area, for the relief of potential development of hydrostatic pressure.

A systematic quality control was applied during the stabilization works, comprising:

- Laboratory testing for the determination of the mortar and the grouting mechanical properties
- Laboratory tests on the rock samples from the sampling boreholes for the investigation of their natural and mechanical properties
- Pull out anchor tests for the assessment of their performance (verification of the design)

The stabilization works begun in November 2008 and successfully completed in December 2008 (Fig. 36a, b).

The documentation of the works comprised a detailed diary, the as built drawings, the quality control report, the technical report of the construction, the geometric (topographic) documentation of the supporting elements and the application of optical fibers sensors in an anchor for continuous monitoring of its stress and strain state (Egglezos 2010a).

5.2.2 *Rescue reburial of Arrephorion*

The remains of the Arrephorion foundation walls in the north side of the Acropolis hill had to be reburied since the weathering effect of the environment was devastating for the sensitive stone blocks

Figure 36. SE area of the rockhill. Before (a) and after (b) the stabilizing interventions.

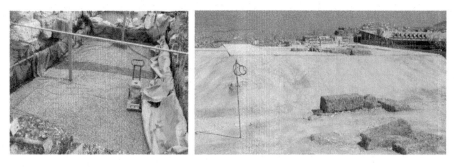

Figure 37. The Arrephorion reburial project during (a) and after (b) the works.

of the walls (of soft marly limestone). The main discussion for the reburial was focused on the architectural features of the completed work, the use or not of the existing material from the ancient artificial filling etc. Although these issues are- of course- an important part for the reburial approach, the main problem was the fact that the North Wall of the monument is in full contact with a part of the Circuit Wall of Acropolis (Fig. 16). The typical reburial for that (north) side could exercise a significant thrust in the (about 5 m high) Circuit Wall (having in mind that there is no interlocking of the Circuit Wall and the foundation walls of the Arrephorion). This self-supporting construction satisfies the criterion for minimization of horizontal thrusts (for both static and seismic conditions). The construction of the reburial was completed successfully in May 2007 (Egglezos 2007) (Fig. 37a, b). Henceforth, the response of the system re-burial-monument is observed systematically through various monitoring applications. The recordings of the monitoring applications are consistent to the expected response from the design and to the results of rigorous 2-D elastoplastic analyses (Egglezos 2013).

5.2.3 *Study of the Parthenon's west side crane*

The study concerns the design of a superficial foundation of a worksite crane (POTAIN type) serving for the restoration of the West Colonnade of the Parthenon (Egglezos 2010b). The main difficulty in the design was the difference of subsoil properties along and across the foundation, the eccentric loading arising from the crane function, the effect of the crane and the foundation loading to the nearby existing part of the Mycenaean fortification wall (Fig. 38).

The foundation has the long axis in the direction N-S and consists of two parallel foundation beams (inverted T), 26 m long and base width B = 0.90 m. Across the beam there are transverse crossbeams every 4.30 m, for the conjunction of the main foundation beams. The height of the main beams is changing from 1.80 to 0.70 m in the north part of the foundation (22 m long) and from 0.70 to 0.96 in the south part (4 m long) due to the inclination of the terrain. It is worth to mention the need for absolutely superficial seating of the foundation on the rockhill surface, since the Acropolis rock itself is considered to have a monumental value. For this reason the concrete foundation was constructed upon a separation geotextile and the system crane—foundation was checked against potential sliding.

The foundation lies partly upon the limestone outcrop of the hill (north part) and partly upon the artificial fill of varying thickness along and across the structure axis (see also Fig. 10). Due to this fact the subsoil conditions of the foundation change continuously. For this reason a detailed discretization of the foundation contact surface was required so as the resulting subgrade modulus (for the structural

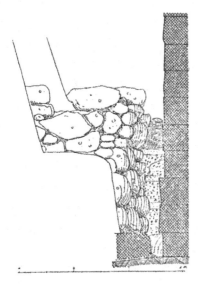

Figure 38. SW corner of Parthenon stereobate at junction with the Mycenean fortification wall. (After Bundgaard 1976, Fig. 32)

Figure 39. The crane after its installation.

Table 6. Geotechnical parameters of the geometerials involved in the study of the Parthenon's crane.

Geotechnical parameters				
	Rockmass	Fill	Added gravel	Concrete
γ (kN/m^3)	26	20	18	25
c (kN/m^2)	250	3–5	0–3	–
φ ($^\circ$)	45	33	36	–
E (MPa)	6000	10	60	16000
σ_{cd} (MPa)	40	–	–	18
$\sigma_{c\text{-}1d}$ (MPa)	2.41	–	–	–
$\sigma_{c\text{-}tx}$ (MPa)	7.61	–	–	–
σ_t kPa	77	–	–	1600
c_u (kN/m^2)	–	–	–	–
ν	0.20	0.30	0.30	0.20

design of the crane foundation) to be representative of the geotechnical conditions. In order to meet the high requirements of the crane designer for low subsidence of the foundation, the fill was excavated and replaced (to a maximum depth of 1 m) by plain concrete and well compacted gravel.

As far as the seismic design of the crane is concerned, pseudostatic analyses were performed. From the analyses resulted that the marginal soil acceleration for the crane overturning ranges from 0.07–0.16 g in relation to the loading state and position of the crane. These limits have been accepted by the responsible authority, taking into account the estimated life time of the crane functioning (<10 yrs).

The geotechnical properties of the geomaterials involved in the foundation design (limestone, artificial fill) were determined on the basis of: a) the existing laboratory data (Andronopoulos and Koukis, 1976, Table 3) and the application of the GSI method for the estimation of the limestone rockmass properties and b) empirical relations for the mechanical properties of the artificial fill based on notional value of in situ SPT test results (suitable for this case).

The artificial fill, according to macroscopic observations and the description from the excavation reports (Cavvadias and Kawerau 1907, Bungaard 1974, 1976) consisted of a soil mixture with skeleton of clayey sand and significant presence of gravel with pebbles (Fig. 38). From a geotechnical point of view, the fill was classified as GC-SC (clayey GRAVEL with sand—clayey SAND with gravel). A rather loose compaction of the fill was taken into account, since the original fill has been replaced with unknown degree of compaction after the completion of the great excavation (1880–1895). For the needs of the design a low relative density (DR = 40%) and a respective low notional SPT value (N_{160} = 10) were assumed. These values are consistent to empirical relation for the relative density: $DR(\%) = 100*(N_d/60)^{0.5} = 40.8$ (Marcuson and Bieganowsky, 1977).

The assignment of certain fill cohesion was justified from the observation that a vertical cut of about 1.5 m was standing stable.

According to empirical relations for $N_{160} = 10$, a shear strength pair of values, c = 3 kPa and $\varphi' = 33^\circ$ and an elastic Young Modulus E = 10 MPa were assumed .

The geotechnical properties of the geomaterials involved in the POTAIN crane design are summarized in Table 6.

The crane installation was successfully completed in 2011 (Fig. 39).

6 DISCUSSION—STUDIES AND INTERVENTIONS NEEDED FOR THE FUTURE

In the framework of a critical review of the studies and works that have been carried out to date, their partial character and lack of a overall conception for addressing the geotechnical issues of the Acropolis site must be noted, without in any way overlooking the contribution made by each separate project in the acquisition of information.

44

Figure 40. General view of the Acropolis from the SW.

There is urgent need, therefore, for an overall management plan, so that the required geotechnical actions will be rationally performed. This plan should comprise the following studies and the works:

- Stabilization of the north part of the rockhill slopes
- Retrofitting of the foundation of the N-W wing of the Propylaia (Pinakotheke)
- Drainage of the Acropolis plateau
- Restoration of the Circuit Wall
- Final display of the surface of the Acropolis plateau, after the completion of the restoration works.

Besides the aforementioned- necessary- geotechnical projects, it should be noted that every intervention on the surface of the Acropolis summit, especially if it is carried out in the proximity the Circuit Wall, should take into consideration the potential impact of the intervention on the Wall and the slopes (e.g. the function of the crane on the SE corner of the Wall serving for the lifting of the worksite materials, Fig. 40).

It is evident that the Acropolis monumental complex, because of the great number of factors involved in an overall approach and analysis, presents a complicated technical and scholarly problem, from the standpoint of confrontation.

Hence, all the works must be supported systematically in every circumstance by various fields of science in the framework of a broad interdisciplinary collaboration.

REFERENCES

Ambraseys, N. 2010. On the long-term seismicity of the city of Athens. *Proc. Academy of Athens* Vol. A.: 81–136.
Ambraseys, N.N. & Jackson, A. 1997. Seismicity and strain in the Gulf of Corinth (Greece) since 1694. *J. Earth. Eng.* 1: 433–474.
Andronopoulos B. & Koukis G. 1976. *Engineering geology study in the Acropolis area.* Athens: Institute of Geology and Mineral Exploration.

Bouras, Ch. 1994. Restoration work on the Parthenon, and changing attitudes towards the conservation of monuments. in Tournikiotis, P. (ed.), *The Parthenon and its Impact in Modern Times*: 310–339. Athens: Melissa.

Bouras, Ch. 2007. Theoretic principles of the interventions on the monuments of the Acropolis. *The Acropolis Restoration News*, 7: 2–5.

Brouskari, M. 1997. *The Monuments of the Acropolis*. Athens: Ministry of Culture.

Bundgaard, J.A. 1974. The Excavation of the Athenian Acropolis 1882–1990. Copenhagen: Gyldendal.

Bundgaard, J.A. 1976. *Parthenon and the Mycenaean City on the Heights*. Copenhagen: National Museum of Denmark.

Cavvadias, P. & Kawerau G. 1907. *Die Ausgrabung der Akropolis vom Jahre 1885 bis zum Jahre 1890*. Athen: The Archaeological Society at Athens.

Egglezos, D. 2007. Protective Filling of Ancient Monuments. The case of the Arrephorion on the Athenian Acropolis, *Acropolis Restoration News*, vol. 7 pp. 19–23.

Egglezos, D. 2010a. The use of modern technological applications for restoring the Circuit Walls of the Acropolis. *The Acropolis Restoration News, Proc. One-day conference "Modern Technologies in the restoration of the Acropolis", Athens, 19 March 2010*. Athens: YSMA.

Egglezos, D. 2010b. Study for the "Transpotation and Installation of the POTAIN crane from the north to the south side of Parthenon"—*Volume of Geostructural Analyses*. YSMA (in greek).

Egglezos, D. & Moullou, D. 2011. Back—analysis sheds light on the history of the Acropolis Wall: The interpretation of a permanent structural failure, *Proc. XVth European Conference on Soil Mechanics & Geotechnical Engineering, Hellenic Society for Soil Mechanics and Geotechnical Engineering, Athens*, Vol. A: 1841–1846.

Egglezos, D. & Toumbakari, E.E. 2011. In-situ tests on the Parthenon columns for the assessment of their foundation condition Proc. XVth European Conference on Soil Mechanics & Geotechnical Engineering, Hellenic Society for Soil Mechanics and Geotechnical Engineering, Athens, Vol. A: 1847–1852.

Egglezos, D. 2013. Re-burial of the Arrephorion on the Athenian Acropolis: an in situ rescue intervention against degradation. *2nd intern. Symp. on Geotechnical Engineering for the preservation of monuments and historic sites, May 30-31*, Napoli.

Galanopoulos, A.G. 1955. Seismic geography of Greece, *Annales Geologique des Pays Helleniques, 6, 83-121 (in greek)*.

Galanopoulos, A.G. 1960. A catalogue of shocks with Io > VI or M > 5 for the years 1801–1958. Athens: Seism. Lab. Univ.

Higgins, M. & Higgins, R. 1996. *A geological companion to Greece and the Aegean*. London: Gerald Duckworh & Co Ltd.

Hoek, E., Carranza-Torres, C. & Corkum, B. 2002. Hoek-Brown criterion–2002 edition, *Proc. NARMS-TAC Conference, Canada, Toronto*, 1:267–273.

Hoek, E. & Diederichs, M.S. 2006. Empirical estimation of rock mass modulus. *International Journal of Rock Mechanics and Mining Sciences*, 43: 203–215.

ICOMOS. 1965. The Venice charter 1964, *II International Congress of Architects and Technicians of Historic Monuments*.

Ioannidou, M. 2003. The structural restoration. In Filetici, M.G. (ed.), *Restoration of the Athenian Acropolis (1975-2003), Quaderni ARCo*: 150–155. Rome: Canzemi.

Ioannidou, M. 2006. Seismic Action on the Acropolis monuments. *The Acropolis Restoration News*, 6: 11–14. Athens: YSMA.

Ioannidou, M. 2007. Principles and methodology of intervention for structural restoration. *Proceedings of the XXI International CIPA Symposium, Athens, 1-6 October*: 376–381. Athens: CIPA.

Ioannidou, M. & Paschalidis, B. 2003. Joining the beams of the Propylaia with titanium reinforcements: a new approach, *Proceedings of the 5th International Meeting for the Restoration of the Acropolis Monuments, 4-6 October 2002*: 291–300. Athens: Ministry of Culture.

Ioannidou, M., Moullou, D. & Egglezos, D. 2008. *The Acropolis of Athens: The restoration project*, Athens: YSMA.

Kakderi, K.G. & Pitilakis, K. 2010. Seismic response of gravity qaywalls and proposal of vulnerability curves based on numerical methods, *6th Pan-Hellenic Conference on Geotechnical and Geo-enviromental engineering, 29 May—1 June*, Volos: TEE (in greek).

Kalogeras, I. & Egglezos, D. 2013. Strong motion record processing for the Athenian Acropolis seismic response assessment. *2nd Intern Symp. on Geotechnical Engineering for the preservation of monuments and historic sites, May 30-31*, Napoli.

Kalogeras, I., Evangelidis, C., Melis, N. & Boukouras, K. 2012. The Athens Acropolis strong motion array, *Geophys. Res. Letters, 14, EGU2012-9523*.

Kalogeras, I., Stavrakakis, G., Melis, N., Loukatos, D. & Boukouras, K. 2010. The deployment of an accelerographic array on the Acropolis hill area: Implementation and future prospects. *The Acropolis Restoration News, Proc. One-day conference "Modern Technologies in the restoration of the Acropolis", Athens, 19 March 2010.* Athens: YSMA.

Korres, M. 1995. Significant structural damages of the Propylaia, Proc. of the 4th International Meeting for the restoration of the *Acropolis Monuments, 27-29 May 1994*, Athens: Ministry of Culture.

Korres, M. 2002. On the North Acropolis Wall. In Stamatopoulou, M. & Yeroulanou, M. (eds.), *Excavating Classical Culture. Recent archaeological discoveries in Greece:.* BAR International Ser. 1031: 179–186. England: Oxford.

Korres, M. 2004. The Pedestals and the Akropolis South Wall. In Stewart, A. (ed.), *Attalos, Athens and the Acropolis*: 242–337. Cambridge: Cambridge University Press.

Marcuson, W. & Bieganowsky W. 1977. The SPT and Relative Density of Coarse Sands. *Journal of Geotechnical Engineering, American Society of Civil Engineers,* 103 (GT11): 1295–1309.

Maridaki, A., Laskaridis, Ch., Tzoumousli, E., Fraghiadaki, E. & Frantzi, G. *The case of Arrephorion. Description and doumentation of the preservation state of the foundation blocks.* Athens: YSMA (in Greek)

Marinos, V., Marinos, P. & Hoek, E. 2005. The geological Strength index: applications and limitations. *Bull. Eng. Geol. Environ,* 64: 55–65.

McNamara, D. E. & Boaz, R.I. 2011. PQLX: A Seismic Data Quality Control System Description, Applications, and Users Manual, *U.S. Geol. Surv. Open-File Rept.* 2010–1292: pp. 52.

McNamara, D.E. & Buland, R.P. 2004. Ambient Noise Levels in the Continental United States. *Bull. Seism. Soc. Am.,* 94 (4): 1517–1527.

Monokrousos, D. 1995. The stabilization works of the Acropolis rockhill. *Proceedings of the 4th International Meeting for the restoration of the Acropolis Monuments, 27-29 May 1994*: 177–181. Athens: Ministry of Culture (in Greek).

Moullou, D. & Mavromati, D. Topographic and Photogrammetric recording of the Acropolis of Athens. *International Archives of Photogrammetry Remote Sensing and Spatial Information Sciences,* 2007, Vol 36–5, C53: 515–520.

Pacific Engineering Analysis. 2005. *NEHRP site class (Vs (30m)) amplification factors from site response simulations*: 1–48.

Rockscience Inc. 2007. *Roclab 1.031 software code.* Toronto.

Rondoyanni, Th., Mettos, A., Galanakis, D. & Georgiou, Ch. 2000. The Athens earthquake of September 7, 1999: its setting and effects. *Ann. Geol. Pays Hellen.* 38(B): 131–144.

Schenkova, Z., Kalogeras, I., Schenk, V., Pichl, R., Kourouzidis, M. & Stavrakakis, G. 2005. *Atlas of isoseismal maps of selected Greek earthquakes (1956-2003).* Athens: Evonymos Ecological Library.

Skarlatoudis, A.A., Papazachos, C.B., Margaris, B.N., Theodulidis, N., Papaioannou, Ch., Kalogeras, I., Scordilis, M.E. & Karakostas, V. 2003. Empirical Peak Ground-Motion Predictive Relations for Shallow earthquakes in Greece. *Bull. Seism. Soc. Am.* 93(6): 2591–2603.

Stevens, G.P. 1946. Architectural Studies Concerning the Acropolis of Athens, *Hesperia* 2: 73–106.

Theoulakis, P., Stefanis, A., Karatasios, I. & Gerogiannis, G. 2009. Preliminary Investigation of Mortars for the cementation of porous (from Piraeus coast) architectural members of the Acropolis Monuments. Laboratory for Conservation of structural Stone, Athens: YSMA (in Greek).

Tsokas, G., Tsourlos, P., Tokmakidis, K., Sarris, A., Stambolidis, A. Papadopoulos, N., Styllas, M. & Naxakis, V. 2006. *Electric tomographies on the Athenian Acropolis' Circuit Wall.* Athens: YSMA (in Greek)

Wyllie, D.C. 1999. *Foundations on Rock.* London: Taylor & Francis.

Vardoulakis, I., Kourkoulis, S., Exadactylos, G. & Razakis A. 2002. Mechanical properties and compatibility of the natural blocks in ancient monuments: the Dionysos marble. In Varti-Mataranga, M. & Katiki, G. (eds), *Proceedings of the Interdisciplinary Workshop "The building stone in monuments", Athens, 9 November 2001*: 187–210. Athens: IGME. (in Greek)

Vardoulakis, I., Kourkoulis, S., Pazis, D. & Andrianopoulos N. 1995. Mechanical behavior of Dionysos marble in direct tension, *Proceedings Felsmechanik Kolloquium*, Karlsruhe.

Zambas, K., Ioannidou, M. & Papanikolaou, A. 1986. The use of titanium reinforcement for the restoration of marble architectural members of the Acropolis monuments. *Proc. IIC Congress on Case Studies in the Conservation of Stone and Wall Buildings*: 138–141. Bologna: International Institute for Conservation of Historic and Artistic Works.

Geotechnics and Heritage – Bilotta, Flora, Lirer & Viggiani (eds)
© 2013 Taylor & Francis Group, London, ISBN 978-1-138-00054-4

The role of geotechnical conditions in the foundation, expansion and preservation of the ancient town of Orvieto (Italy)

P. Tommasi
CNR—Institute for Environmental Geology and Geo-Engineering, Rome, Italy

M. Sciotti
Department of Civil and Environmental Engineering, University of Roma 'Sapienza', Italy

T. Rotonda & L. Verrucci
Department of Structural and Geotechnical Engineering, University of Roma 'Sapienza', Italy

D. Boldini
Department of Civil, Chemical, Environmental and Material Engineering, University of Bologna, Italy

ABSTRACT: Orvieto rises on top of a mesa formed by a slab of weak pyroclastic rocks overlying an overconsolidated clay formation. The particular geotechnical conditions of the site enhanced the erosive process that isolated the mesa and in turn favoured human settlement. These geotechnical conditions include the stiffness contrast between the slab and the clay, the low strength of the pyroclastics, and the susceptibility to degradation of the clay formation. These conditions have been found to be the source of a complex and time-dependent interaction between landslides and urbanization. In this paper the anthropic changes to the natural evolution of the landscape are reconstructed through archival and historic documents over the last eight centuries and interpreted from a geotechnical point of view. The types and characteristics of failures that occurred in the past are described by means of aerial photo analysis, field surveys and archival research, whilst geometry and kinematics of active movements are illustrated using present-day monitoring results. Based on a detailed interpretation of displacement and pore pressure monitoring in the clay slope and the results of numerical analysis of the erosional process and selected failures, a conceptual model of the instability mechanism of the hill is proposed. Finally, protection policies from the Renaissance period until the XX century, when comprehensive preservation projects started, are illustrated and their impact on the instability mechanism discussed.

1 INTRODUCTION

Most of the towns of Western-Central Italy, corresponding to the southern part of the ancient Etruria region, originated as Etruscan centres that not only survived but also expanded throughout the Roman and Medieval periods.

Many of these towns rise on flat-topped hills (buttes and mesas) that are delimited by steep escarpments and totally isolated from the surrounding rock plateau. The upper part of the stratigraphy consists of competent soft or hard rock for the first few tens of metres, whereas the lower part of the slope consists of stiff Plio-Pleistocene clays.

These geotechnical conditions played a decisive role in the foundation of these towns because they enhanced the erosional processes that isolated the hills from the surrounding rock plateau, thus making them preferred sites for early human settlement due to the natural defence that they provided against enemy attacks.

On the other hand, these same geotechnical conditions make the rock slab margin instable, thus influencing urban development and preconditioning the remedial measures chosen to preserve these historic sites from the progressive retreat of the rock cliff. Even though human activity has in some cases modified the erosional rate, the instability phenomena are essentially due to the natural evolution of this type of landscape and they cannot be completely stopped in the long term.

For its prominent historical role and its outstanding artistic heritage, the most representative town of this region is Orvieto. Due to its significance, Orvieto has undergone various major re-medial interventions and detailed geotechnical investigations in an effort to preserve this town for future generations. The historical importance of this town throughout the centuries has left an impressive amount of documental/archival data whose analysis has allowed us to reconstruct, from a geotechnical perspective, the interaction between the urban centre and the surrounding environment as well as its progressive changes up to the present.

2   GEOLOGICAL SETTING OF THE ORVIETO HILL

Orvieto rises on top of the largest mesa carved in the volcanic plateau of Central Italy (Fig. 1). The mesa is formed by a tuffaceous slab delimited by vertical cliffs up to 50 m high, overlying a clay base

Figure 1.   View of Orvieto from the South-West.

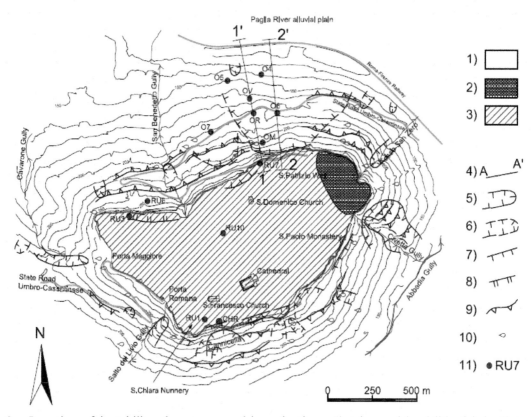

Figure 2.   Location of instability phenomena and investigations: 1) talus and landslide debris; 2) travertine; 3) pyroclastics; 4) trace of geological/geotechnical profile; 5) recent landslide scarp on the clay slope; 6) old landslide scarp on the clay slope; 7) landslide scarp at the slab margin; 8) limit of lowered block at the slab margin; 9) landslide terrace on the clay slope; 10) spring; 11) borehole.

50

with gentle slopes (11° to 16°). The slope is covered by talus and slide debris and is furrowed by several gullies which originate at the foot of the cliff and flow radially along the clay slopes (Fig. 2).

A 2- to 15 m-thick succession of fluvio-lacustrine sediments (sands, gravels, diatomites and pumice layers) is interposed between the pyroclastic slab and the clay formation. This succession hosts a perched aquifer that seeps out at the cliff foot as a series of perennial or intermittent springs.

3 THE GROWTH OF THE URBAN CENTER

Even though the first permanent human settlement dates back to the IX century B.C., the Etruscan town experienced progressive economic and demographic growth from the VII to III century B.C., after which it was destroyed in 264 B.C. by the Roman Republic and the surviving inhabitants moved close to Bolsena Lake.

Evidence from archaeological excavations indicate that the Etruscan town extended over the western third of the rock slab, where the only natural access to the hilltop was located. Two large necropolis (*Cannicella* and *Crocifisso del Tufo*) were located at the base of the southern and northern cliffs, respectively. Tangible traces of the Roman period have not been found and the hill top is believed to have been abandoned until the VII century.

In keeping with the typical piecemeal and organic medieval approach to urbanization, the town adapted to topographic conditions: the butte edges defined the limits and shape of the town and the rock cliffs replaced walls and fortifications, providing the building stones for construction of most of the public, religious and residential buildings.

Urban expansion occurred mainly across the XIII and XIV centuries, at the apex of the communal government, when Orvieto had a population of about 20,000 inhabitants (Fig. 3). Major private residences and public/religious edifices/infrastructures were built during this period. Major works included the aqueduct (whose terminus is excavated into the tuff and which still feeds the public fountains on the mesa top), the four town gates, and several churches. The Cathedral (one of the outstanding edifices of medieval architecture in Italy) and the churches of S. Francesco and S. Domenico are all imposing and structurally complex buildings. Monasteries and nunneries were built at the margin of the urban area, close to the cliff edge.

Subsequent internal conflicts between supporters of the emperor and the pope (Guelphs and Ghibellines) marked a stop in building activity and a decline in the population.

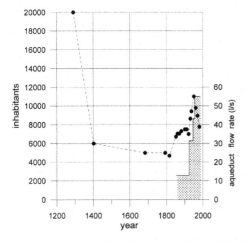

Figure 3. Population (dots and dashed line) and water consumption (hatched bar graph) vs. time (data from ISTAT and Perali 1919).

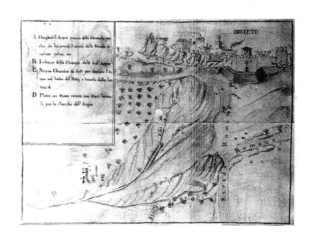

Figure 4. Watercolour depicting the *Cavarone* gully and the failure which involved the road to Rome before 1751 (State Archive of Orvieto).

The fabric and extension of the medieval urban area remained virtually unaltered until the XVII century. At that time Orvieto extended over most of the present day constructed area, except for the easternmost part of the hill top, which was subsequently developed over the following centuries.

In the XIX century, construction and integration of edifices in the urban fabric followed the criteria adopted in the previous centuries, whilst those made during the XX century are in sharp contrast with the older architecture and drastically break the visual and structural continuity of the town.

4 IMPACT OF HUMAN ACTIVITIES ON GEOTECHNICAL CONDITIONS

Two main anthropic activities, linked to the growth of the urban centre over the centuries, have altered the geotechnical situation and influenced stability conditions:

– progressive changes to the surface water and groundwater regimes;
– underground excavations in the pyroclastic formation, which locally led to the collapse of marginally stable parts of the rock mass.

4.1 Changes to hydrology and hydrogeology

Street paving and sewer system construction started in the XIX century. Rain/waste waters were thus collected and discharged through wells at the foot of the cliff, where gullies convey them towards the main water courses. In some areas rain and waste waters have been directly discharged onto the clay slope.

Street pavement and the consequent catchment of rainwater progressively reduced infiltration on the slab top, whilst areas of uncontrolled discharge at the cliff edge greatly altered local piezometric conditions. In addition, controlled localized discharge has caused increased erosional activity along the gullies, which has triggered failures on their banks and in the adjacent areas.

Historical evidence of erosion and slope failure on the gully banks have been found in archives. For example, the XVIII century watercolour in Figure 4 depicts the failure of the road to Rome due to erosion along the *Cavarone* gully, originating at the north-eastern margin of the rock slab (Tommasi et al., 1986). It also depicts the discharge of waste water on the clay slope at the time of failure (including a check dam built for reducing water energy) and the hydraulic works planned to move the buried conduit uphill and farther from the road.

Figure 5. a) Detail of Sanvitani's 1662 engraving of the southern cliff (State Archive of Orvieto). The *Salto del Livio* gully, originating at the foot of the cliff, is just visible upslope from the cultivated field; b) photograph taken before 1900 where the *Salto del Livio* gully deeply furrows the slope.

Erosion along the gullies varied over the clay slope. In the XVIII century the head of the *Cavarone* gully was already sharply incised (Fig. 4), whilst the *Salto del Livio* gully was only just visible in a vivid view of the southern cliff engraved by Antonio Sanvitani in 1662 (Fig. 5a). The same gully, at the end of the XIX century, looked deep and bordered by steep and ravined banks (Fig. 5b). The *Civetta* gully (south-western part of the hill) showed a similar aspect at the beginning of the XX century, when the discharge of waste water was not regulated and banks showed signs of apparent instability (Fig. 6).

4.2 *Underground and surface excavations*

Since the Etruscan period, the pyroclastic slab has been excavated for building material (pozzolana for mortars and lithic tuff for masonry blocks), storage of goods/water, and other minor activities (e.g. pigeon keeping). More than 500 cavities were mapped during the conservation program started in the 80's (see section 9).

Exploitation of pozzolana and tuff reached its peak in the Middle Ages and then maintained the same production levels until the XVII century, after which the demand gradually decreased. In particular construction from the XIII to XVII centuries depended primarily on tuff blocks for all types of buildings. Surface exploitation started at the western margin of the slab, named *cava* (quarry in Italian) because of a sharp pit morphology which could not have formed from only natural morphological evolution (Stiny, 1930).

Since the Middle Ages building materials have been excavated under the public authority, which regulated exploitation through specific mining claims. This practice was particularly well-documented for the construction of public edifices. Tuff or pozzolana failures and marginally stable tuff blocks were occasions for quarrying. The two practices, that had been occasionally monitored in the past, were forbidden. Furthermore order was given to remodel the land surface at safe slope angles and to reforest debris bodies of recent failures. Similarly, brick lining and in-filling (with materials "stronger than the tuff") of cavities excavated underneath buildings was prescribed.

A chronicle from 1511 reports the death of a quarryman due to a vault collapse in a pozzolana quarry indicates that safety conditions in quarries have been a significant problem since the Renaissance. Monitoring of quarrying activity was necessary not only for preventing casualties but also to prevent larger consequences, such as the instability of the cliff and the buildings above the under-

Figure 6. Head of the *Civetta* gully at the beginning of the XX century. A slump on the left and the waste water fall at the centre of the photograph are visible.

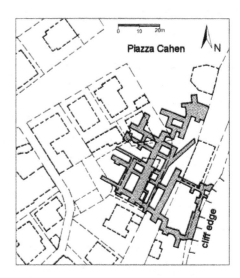

Figure 7. Map of the room-and-pillar quarry of Piazza Cahen (in the eastern part of the town) excavated in the upper pozzolana layers of the mesa. Building footprints are reported with a dashed line.

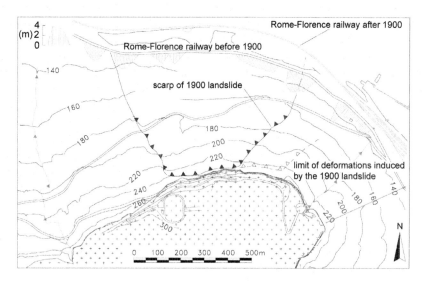

Figure 8. Location and height of the cuts carried out in 1865 for the construction of the Rome-Florence railway (data from Ferrovie Meridionali 1888).

ground quarries. The latter issue became important only when the town expanded towards the eastern part of the slab where the larger quarries were located (Fig. 7). An early emblematic Municipal Board document drawn after a cliff failure in 1583 prescribed quarry inspections by technicians and inhibits excavation under public and private areas.

Since the second half of the XIX century the Municipal authorities have been much more severe, probably as a consequence of the repeated failures that occurred on the southern cliff and which required direct intervention by the national bureau of mines. For example, in 1893 several quarries were closed while others were forced to undertake stabilization work or limit themselves to a safe distance from buildings and infrastructure (De Marchi, 1897).

In 1897 the commission in charge of investigating structural integrity of the town forbade all quarrying activities and prescribed monitoring of all excavations. In actual fact quarrying did continue after this edict but at a much reduced level, with a few quarries remaining active until the 1940's in scarcely inhabited areas (Fig. 7). Instead, utilization of underground cavities for reasons other than exploitation continued until the 1980's.

Modifications to the geometry of the rock cliff and the clay slope are less well documented but important. A particularly relevant case is represented by the excavations carried out for the construction of the Rome-Florence railway, terminated in 1865. Railway cuts, whose location and height are reported in Figure 8 were considered to have triggered the 1900 *Porta Cassia* landslide. The relationship between excavation and failure is discussed in section 7.

5 INSTABILITY PHENOMENA AND REMEDIAL MEASURES THROUGH THE CENTURIES

5.1 *Instability phenomena*

A landslide catalogue was firstly drawn by Tommasi et al. (1986) and successively integrated by Martini & Margottini (2000).

As soon as the first newspapers appeared (1280), landslides were noted and even described. At that time (see section 3) large parts of the town had already been built close to the cliff edge. Subsequently landslides appeared in the official acts of the local and central government. Until the XVIII century the documented landslides involved only the cliff, which always recurred in the same areas over time. Cliff failures directly threatened buildings at the slab margin or the road that wound along the cliff base. Landslides on the clay slope have been reported with continuity since the slope became permanently inhabited (i.e. the end of the XIX century), unless they involved infrastructure. Therefore the types of documented landslides progressively increased with time until, at the beginning of the XX century, they encompassed all observed instability phenomena, including failures on the cliff and shallow translational landslides as well as slump failures of the gully banks on the clay slope.

Historical data also document two large failures, one in 1898 or 1899 on the cliff underneath the S. Chiara nunnery (a large sector of tuff and pozzolana) and another in 1900 on the northern clay slope (the *Porta Cassia* landslide), which have never re-occurred. The 1899 collapse (Fig. 9) represented the apex of a sequence of failures, started in 1886 and interrupted in 1919, which changed the cliff landscape and damaged the overlooking buildings. The progressive disruption of the S. Chiara cliff was portrayed by photographers and meticulously described by technicians of the Orvieto Municipality, who recorded more than 15 failures between 1886 and 1903. Vinassa de Regny (1904) estimated an overall retreat of the cliff edge between 1835 and 1903 of about 30 meters. By merging photographic documents and the abovementioned report, rock failures were intercalated with, at least, two landslides in the clay slope (the earlier one indicated in Figure 9 and another in 1900), thus highlighting a striking interdependence between rock and clay failures.

A similar process, but in the form of large deformations, between the 1970's and 1980's contributed to the development of a large-scale protection programme for the entire hill.

5.2 *Remedial measures until 1979*

When the urban centre started to expand towards the peripheral areas, deformations and failures at the cliff margin could no longer be ignored and work was undertaken to stabilize unstable areas and to mitigate landslide risk. Over time the cliff and underlying slope have been the object of remedial measures, including:

– protection and support of the cliff walls;
– control of water discharge along the gullies (check dams, protection of gully banks);
– slope remodelling and catchment of surface water.

These interventions were performed over the entire hill, however over time their intensity, mode of action, and relative importance have significantly changed.

Figure 9. Photograph taken in 1900 showing the collapse of a large sector of the tuff slab in 1898 (1899) underneath the S. Chiara nunnery. The arrows and the dashed line indicate, respectively, the crown scarp and the body of a previous landslide on the clay slope.

Until 1936, actions undertaken by the Pontifical State, and successively by the Orvieto municipality, were aimed to solve local instability situations.

The first notices of remedial measures date back to the XVI century, when the Pontifical State had at its service outstanding architects like Antonio da Sangallo, and refer to retaining and lining walls protecting the rock cliffs. This typology represents most of the documented stabilization works until the second half of the XIX.

Figure 10a shows the reconstruction, based on period sources, of a 18-m-high counterfort retaining wall envisaged to stop the repeated failures along the southern cliff. The length of the retaining wall that was actually built was much reduced. Figure 10b illustrates the failure of an older counterfort wall (1660) and its geometry.

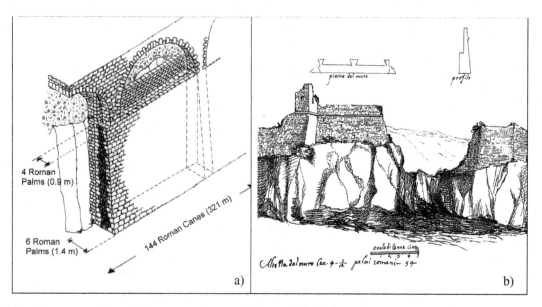

Figure 10.   Reconstruction of the counterfort retaining wall built in 1708 at *Ripa Medici* (*Medici* cliff), based on a report of the Pontifical State technicians a), and geometry of a 24 m long retaining wall collapsed in 1660, sketched by Pontifical State's technicians b) (State Archive of Rome).

Figure 11.   View of the south-western zone (*Foro Boario*) during the interventions carried out at the end of the XIX century.

Only two interventions on the clay slope are well documented prior to 1870: the canalization of the waste waters in the western part of the hill (Fig. 4) and the remodelling and regulation of surface water circulation in the south-western zone (*Foro Boario*). The latter intervention is visible in Figure 11.

The promulgation in 1937 of a Royal Decree, that included Orvieto on a list of urban centres to be protected with funds of the Italian State, represented a change in the intervention practice. This act, expression of a wider change in the public works policy of the Fascist regime, resulted in a detailed programme of integrated measures for stabilization and risk mitigation, which envisaged exceptional funding for that time (Tommasi et al., 1986). The programme was never actuated during the regime due to the dramatic effects of the Second World War. Part of the stabilization and protection works was progressively carried out in the following three decades, but the comprehensive approach was lost and Orvieto had to wait 43 years to have another similar chance (see section 9).

6 GEOTECHNICAL CHARACTERIZATION OF THE MATERIALS

6.1 *The pyroclastic slab*

The pyroclastic slab is composed of two main facies: a lithic tuff, affected by marked zeolitization processes, and a weakly cemented tuff (pozzolana). Other minor lithotypes can be identified through a more refined analysis. For example, a yellow and less resistant tuff can be distinguished from the more common red tuff.

The pozzolana facies can also be differentiated into at least two sub-types, based on strength and abundance of hard pumices. Their different resistance to erosion influences the morphology of the cliff, which is gentle in the weakest cap layer (where the largest cavities were excavated), steeper in the under-lying stronger lower pozzolana, and vertical in the basal tuff layer. The passage between the materials can either be gradual or form horizontal discontinuities.

The rock mass structure is clearly visible beneath the S. Francesco Church (Fig. 12). Here the tuff is cut into prisms by systematic vertical discontinuities that extend upwards into the hard pozzolana. Two sets of sub-vertical discontinuities have been found, one parallel (K_1) and the other orthogonal (K_2) to the cliff edge. Two other sub-vertical conjugated sets, K_3 and K_4, form an angle of 45° with the cliff face (Manfredini et al., 1980). These are typical of spurs and edges, where they create a "columnar" struc-ture. Discontinuities usually have small apertures (a few millimetres) and are quite persistent. These characteristics are related to the state of stress at the slab margin.

Figure 12.   View of the southern cliff.

57

In contrast the overlying weak pozzolana layer is massive with irregular and less persistent discontinuities (Fig. 12), similar to that observed in other areas (i.e. S. Bernardino monastery). Surveys in two large cavities (n° 536 and 508) indicate that discontinuities in this unit are typically sub-vertical, non-persistent, undulated, closed and sometimes slightly oxidized (Tommasi et al., 2006b). They are very widely spaced (mean spacing of about 3 m). The survey of cavity n° 508 confirms the presence of K_1 and K_2 sets (Fig. 13). The dispersion in orientation increases in cavern n° 536 (highly branched), probably due to the presence of K_3 and K_4 sets, although discontinuities due to excavation-induced stresses could overlap the pre-existing ones.

The in situ mechanical characteristics of the different layers in the S. Francesco area were investigated using dynamic surveys. A cross-hole test (Rotonda et al., 2002) was carried out 5 m behind the cliff face (Fig. 12) through the entire pozzolana sequence down to a depth of 36 m (Fig. 14).

A SASW survey was carried out in the n° 508 pozzolana quarry, extending perpendicular to the cliff wall at 11 m below the surface and 50 m besides the cross-hole test site (Fig. 12). The investigated depth did not exceeded 7 m, due to the poor elastic characteristics of the pozzolana and to the impossibility of utilizing low-frequency and high-energy sources. The S-wave velocity (Fig. 14), below a very soft 3 m thick layer due to excavation disturbance and to floor loosening, gradually increases with depth and is consistent with CH results.

At the laboratory scale the pyroclastic materials of the Orvieto slab have a similar composition, consisting of pumice fragments of different sizes (from a few millimetres to few centimetres) embedded in a glassy ground mass, with rarer phenocrystals and lithic fragments. Two main facies can be differentiated based primarily on textural continuity of the ground mass and mineralogical composition: a lithic facies (tuff) and a poorly cemented one (pozzolana). SEM and X-ray diffraction analyses reveal that the tuff is characterized by a high content of void-filling zeolite minerals that contribute to the higher

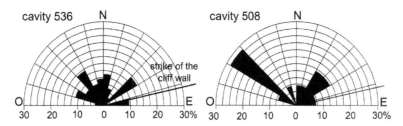

Figure 13.   Rose diagram of joint strikes surveyed within two cavities.

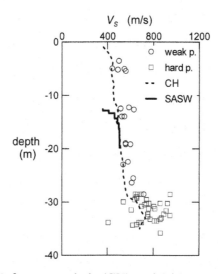

Figure 14.   Shear wave velocity V_S from cross-hole (CH) and SASW survey and from laboratory tests (dry conditions) on pozzolana specimens recovered in the borehole utilized for the CH test.

cementation (Tommasi & Ribacchi 1998). Zeolites are absent in the pozzolana, which also seems to have a less continuous glassy ground mass.

The pyroclastic materials have a very high porosity (from 45 to 60%) and, thus, a low density (dry density γ_d ranges from 1.00 to 1.25 Mg/m³). The low density of the lithic tuff is a result not only of the widespread macro- and micro-pores but also of the high content of zeolite minerals having a low grain density (2.1 Mg/m³ versus 2.7 Mg/m³ of the volcanic glass forming the pumices and the ground mass). Porosity mainly influences dynamic properties (Fig. 15) and strength (Fig. 16) of the various materials.

The dynamic properties of the pyroclastic materials cover a wide range, such as V_S values from 400 m/s in the weaker pozzolana up to 1200 m/s in the red tuff. The Poisson ratio, v, varies between 0.2 and 0.4. A strict agreement between laboratory (dry conditions) and site measurements of the V_S has been found for the pozzolana deposit, as confirmed by cross hole and SASW tests (Fig. 14). Based on these results it can be inferred that the discontinuities exert a very weak influence on the dynamic behaviour of the rock mass.

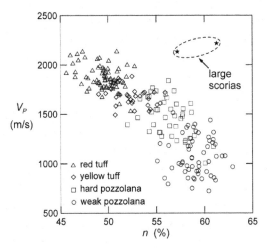

Figure 15. Velocity of longitudinal waves V_P in the pyroclastic materials versus porosity n.

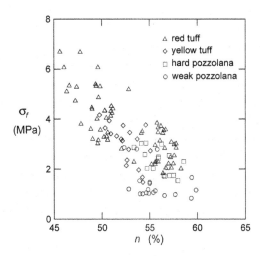

Figure 16. Uniaxial compressive strength σ_f of the pyroclastic materials versus porosity n.

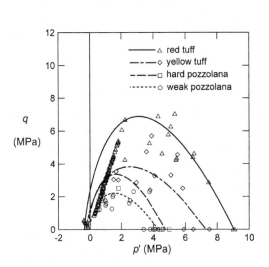

Figure 17. Yielding stress from triaxial tests in the pyroclastic materials.

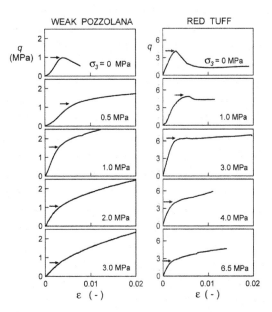

Figure 18. Deviatoric stress q vs. axial strain ε in triaxial compression tests at increasing confinement stress σ_3. Yielding is highlighted by arrows.

The lower value of the measured uniaxial compression strength σ_f and tensile strength σ_t are 0.8 and 0.1 MPa, respectively.

All facies yield at low isotropic stress values, similar to many other weak rocks (Fig. 17). In particular the weakest pozzolana facies yields at a mean effective stress $p'_y = 4$ MPa, while the red tuff yields at a pressure of 6–9 MPa.

The pozzolana exhibits a strain hardening behaviour even for moderate confining stresses (> 0.5 MPa) (Fig. 18). Conversely the tuff holds a strain-softening behaviour up to $\sigma_3 = 3.0$ MPa, i.e. about 40% of p'_y.

6.2 *The clay slope*

The clay slope is formed, from top to bottom, by talus debris, landslide debris, and stiff clay, and has an upper softened portion (Figs. 19 and 20). Moving downslope from the cliff foot, the talus debris passes from tuff blocks immersed in a volcaniclastic matrix (thickness up to 7–8 m) to a mixture of pyroclastic grains and clay including tuff blocks and remnants of man-made structures (thickness ranges from less than 1 m to 4–5 m).

The landslide debris (2–10 m thick) is an aggregate of oxidized clay peds, derived from the remoulding of the clay formation by slope movements. The contact with the underlying clay formation is often represented by a sharp surface.

The clay formation at depth is stiff, without apparent discontinuities. It is gray-blue in colour and looks homogeneous with interspersed, carbonized vegetal remnants and small iron oxide nodules. Even though it is stiff, the uppermost interval (up to 12 m thick) is softened (Fig. 20) and characterized by tight fissures and regularly-spaced sub-vertical joints caused by slope deformations, both with oxidized faces. Proceeding downwards discontinuity frequency gradually decreases and softening is revealed only by changes in index properties (Fig. 20).

The intact clay and its uppermost softened layer can be classified as clay and silt of medium plasticity (Tab. 1). Calcium carbonate (30–35%), in the form of microfossils, does not bond the clay aggregates. Nevertheless, standard oedometer tests (up to an effective vertical stress of $\sigma'_v = 8$ MPa) do not reveal a well-defined preconsolidation stress, which can be assumed to be no lower than 1.8 MPa from the geological history of the hill. According to their consistency index ($I_c = 1.20$), the stiff and softened materials lay in the semisolid state.

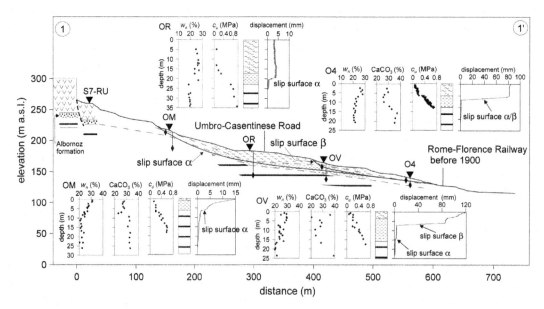

Figure 19. Geotechnical profile (Section 1–1′ in Fig. 2) with deep (α surface) and shallow (γ and β surfaces) movements detected by inclinometers. Plots of displacements vs. depth refer to a two-year period.

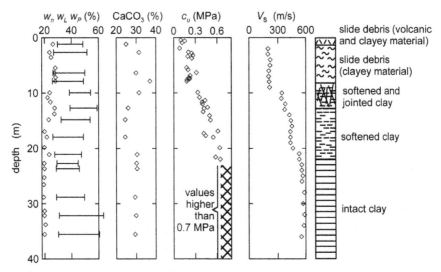

Figure 20. Typical log of the physical-mechanical properties of the clay slope materials.

Table 1. Index properties of the clay materials (natural state).

Material and sample	γ (kN/m³)	w_l (%)	w_p (%)	w_n (%)	e	c_u/σ'_{v0} (kPa)
Intact	20.6	63.0	29.0	20.0	0.54	1.32
Softened (S1)	20.5	50.1	28.4	23.5	0.64	2.60
» (S2)	20.3	54.5	30.3	23.0	0.62	1.95
» (S3)	19.5	45.1	26.4	27.0	0.73	0.79
» (S4)	–	52.3	27.6	30.0	0.81	–
» (S5)	19.4	69.3	33.0	27.3	0.74	0.75
Remoulded	19.1	54.0	25.0	26.0	0.70	0.91

γ: bulk unit weight; w_L: liquidity index; w_p: plasticity index; w_n: water content; e: void ratio; c_u: undrained shear strength.

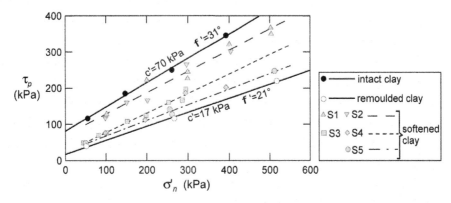

Figure 21. Peak shear strength from direct shear tests on the clay materials. The strength envelopes of the intact clay (black circles) and the remoulded clay forming the landslide debris (empty circles) delimit the strength range covered by materials with different degrees of softening.

Despite the uniformity of the clay formation, the drained peak shear strength, measured using direct shear tests on undisturbed softened clay samples, depends on the stress history and "physical environment" at the different sampling sites (Fig. 21) (Tommasi et al., 1997).

Tests on a block sample retrieved from a fresh cut along the *Civetta* gully (S1 in Figs. 21–22) and on a sample recovered in a borehole drilled at the cliff foot (S2 in Figs. 21–22) indicate that when the clay has only experienced mechanical unloading its strength parameters are close to those of the intact material ($c' = 70$ kPa and $\phi' = 31°$).

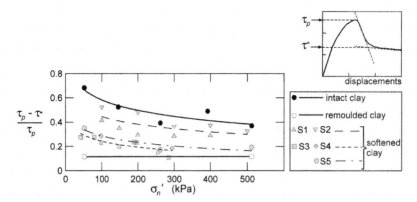

Figure 22. Strength drop after peak ($\tau_p - \tau^*$) normalised to the peak shear stress τ_p vs. normal stress (from direct shear tests).

The failure envelope drops dramatically when materials have experienced pronounced swelling, due to a complete removal of confinement stresses coupled to prolonged soaking. This is the case of the block sample recovered at the foot of the right bank of the *Civetta* gully (S3 in Figs. 21–22) and of the specimens directly recovered from the clay exposed on the *Cannicella* landslide scarp (S4). This was located at the cliff foot where abandoned sewers continue to discharge water on the slope surface.

Test results on the sample recovered within the softened clay layer in the OR borehole drilled into the 1900 *Porta Cassia* slide mass (S5 in Figs. 21–22) require a different analysis. The strength envelope ($c' = 25$ kPa, $\phi' = 23°$) lies unexpectedly close to that of the completely remoulded material ($c' = 17$ kPa, $\phi' = 21°$). Considering the remarkable sampling depth (17–20 m) such a strength degradation is likely due to the intense shearing experienced by the clay during various sliding events (with the 1900 *Porta Cassia* landslide being the last one) (see section 7).

Values of natural water content (27%) and bulk unit weight (19.4 kN/m³), significantly lower than those typical of overconsolidated clays, imply that the uppermost part of the clay formation in the *Porta Cassia* area has been subjected to intense degradation.

Conversely, the relatively small variability of the drained residual strength ($\phi'_R = 15.6°$) highlights the intrinsic uniformity of the clay material.

The slope materials also have different permeabilities. Falling head tests in the Casagrande piezometers yielded permeability values of 10^{-11}, 10^{-10}, and 10^{-9} m/s for the stiff clay, softened clay, and slide debris, respectively (Tommasi et al., 2006a). An increase in permeability is to be expected in the softened and remoulded clay due to fissures and joints, whilst the stiff clay can be more conductive locally when thin sandy laminae are present.

7 GEOTECHNICAL ANALYSIS OF THE INSTABILITY PHENOMENA

7.1 *The clay slope*

The instability phenomena which affect the clay slope (Fig. 2) are characterized by different mechanisms, as the mechanical properties, geotechnical stratigraphy and anthropic actions have a different relative importance. The most widespread form of instability consists of slow translational intermittent movements within the over-consolidated clays at depth or within the overlying debris cover.

Shallow local failures within the debris cover have also been frequently recorded in the past over the entire slope, whilst deep failures in the clay formation represent isolated events that at present seem to be strictly related to anthropic activities.

7.1.1 *Types of failures*

The most frequent type of failure observed before implementation of the protection programme was shallow slumping on the steep gully banks. Failure occurred when the erosional activity within the gullies, documented in section 4.1, increased the bank height until it reached values that were not compatible with the strength of the landslide debris or of the intensely softened clay exposed at the bottom of the gully (see section 6.2).

Shallow landslides parallel to the slope dip direction have also been recorded. They involved part or the entire thickness of the landslide debris cover. Most of these landslides appear to be old slow movements that were suddenly re-activated by local triggers. In particular, gully erosion favoured landslides close to the gullies, whilst undrained loading induced by large cliff failures triggered landslides in the upper part of the clay slope. The *Cannicella* landslide (1900), which involved the area on the left in Figure 9, occurred after a series of large failures on the rock cliff.

Of the failures that have occurred over the last 110 years, only two have resulted from movement along deep slip surfaces. Both have been connected to excavations or abrupt changes to the local hydraulic conditions.

The *Porta Cassia* landslide (1900) occurred 35 years after a 4 m high, and more than 600 m long, excavation had been made at the slope foot for the Rome-Florence railway line (Fig. 8). The small height of the cut and the low gradient of the slip surface suggest that the landslide occurred, for most of its extension, along a pre-existing slip surface. Only a relatively small area at the slope foot experienced a first-time delayed-failure mechanism (see e.g. Chandler, 1984).

Conversely, the 1979 *Cannicella* landslide occurred along a sharply curvilinear slip surface in a steeper area (22°) and was considered to be a first time slide that was triggered by a progressive increase in piezometric levels due to the uncontrolled intense discharge of waste water from the top of the cliff onto the clay slope (Lembo-Fazio et al., 1984).

Deep failures within the softened clay formation occurred in the past, when morphology was steeper and the alluvial plain was located at a lower elevation. This is directly testified by deep slickensided surfaces observed in borehole cores and large masses of stiff clay found within the alluvial coarse sediments of the alluvial plain, close to the slope foot. Indirect support is also given by limit equilibrium analyses, as discussed in section 8.

7.1.2 *Slow movements and slope monitoring*

Slow movements have been studied by monitoring displacements and pore pressures since 1982 using seven stations installed in a sample area on the northern flank of the hill (Fig. 2). The monitoring system was set up by the National Research Council (IGAG) and Sapienza University (DISG and DICEA).

Each station consists of two boreholes equipped with an inclinometric tube and Casagrande piezometers, respectively. At three locations a third borehole equipped with vibrating wire piezometric cells (continuous reading) was added in 2008. Rainfall and temperature are recorded at the slab top by a meteorological station of the Regione Umbria (Idrografico Regionale).

Deep movements occur at a displacement rate of 2 to 6 mm/year along pre-existing slip surfaces/shear bands located within the softened part of the clay formation (surface γ in Fig. 23). Movements, predominantly translational, involve large areas and are re-activated seasonally.

Shallow sliding of the landslide debris cover is also superimposed on the deep movements (Figs. 19 and 23). These shallow movements, generally translational, have higher velocities and reactivate more than once in a hydrologic year with a maximum displacement rates of 7 to 12 mm/month.

The analysis of displacements, pore pressures and cumulative rainfall indicates that reactivation of both shallow and deep movements is rainfall-induced (Tommasi et al., 2006a).

For the shallow movement (γ slip surface in Fig. 23), variations in pore pressure are well related to rainfall over periods of 15–30 days; movements reactivate only when pore pressure exceeds a threshold (Fig. 24b), as identified through the continuous pore pressure monitoring system. Time histories also show that movements stop for small decrements in pore pressure, suggesting that the slide is close to limit equilibrium conditions.

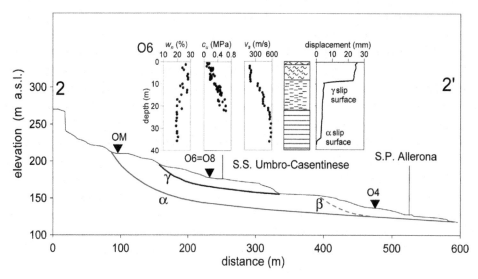

Figure 23. Shallow movement along the γ slip surface detected in O6 inclinometer (trace of Section 2–2' is reported in Figure 2) (modified after Tommasi et al., 2013).

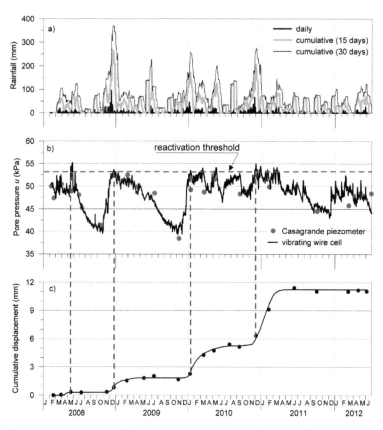

Figure 24. Rainfall a), pore pressure (Casagrande piezometers and vibrating wire cells) b) and cumulative displacements c) along the γ slip surface reported in Section 2–2' (Figure 23) (modified after Tommasi et al., 2013).

For deep movements (α slip surface in Fig. 19) reactivations are related to rainfall accumulations over much longer periods (90 to 120 days) (Fig. 25c) and occur following pore pressure variations of less than 3 kPa (Fig. 25a).

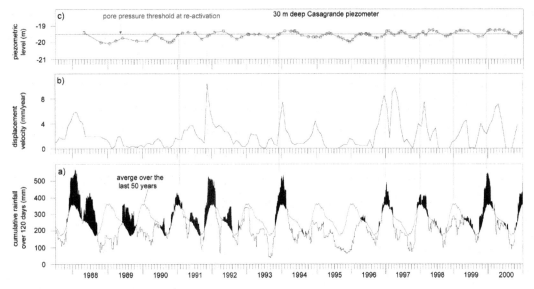

Figure 25. Rainfall a), pore pressure (Casagrande piezometers) b), cumulative displacements c) and displacement rate along the α slip surface (borehole OR in Fig. 19).

Figure 26. Step-like morphology of marginal blocks (S. Chiara). Arrows indicate shear fractures at the base of tuff columns due to basal shear mechanism.

Figure 27. Hourglass-shaped pillar in the pozzolana quarry at the bottom of the pyroclastic slab, upslope from the *Civetta* gully.

In both cases, limit equilibrium analyses have indicated that the shear strength angle mobilized at the limit equilibrium (considering residual conditions) is close to the upper limit of the friction angle measured in the laboratory.

7.2 *The rock cliff*

Cliff failures, the ultimate and most apparent stage of a deeper instability phenomenon, are caused by the interaction between the rock slab and the underlying clay, inducing pervasive fracturing of the lower part of the slab margin. All around the slab, thick slices are lowered with respect to the rock mass behind. The phenomenon involves both columns and thick slices (up to 10 m), which often gives the slab margin a step-like morphology (Fig. 26).

Where the rock consists of lithic tuff with systematic subvertical fractures, failures occur in shear at the base of tall columns subjected to uniaxial compression under its own-weight (in the form of topples), or in tension when weak horizons are intercalated within the tuff. This phenomenon has also been

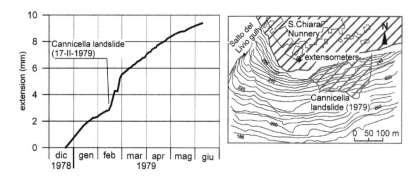

Figure 28. Displacement vs. time recorded by a rod extensometer installed in the S. Chiara spur and location of the extensometer borehole with respect to the 1979 *Cannicella* landslide.

Table 2. Main earthquakes which caused damage at Orvieto.

Year	Epicenter		D (km)	I_0 (MCS)	I_S (MCS)
1695	Bagnoregio	NF	10	IX	VIII
1328	Norcia	FF		X	VI
1703	Norcia	FF		X	VI–VII
1751	Gualdo Tadino	FF		X	VI
1349	L'Aquila	FF	140	X	VII–VIII
1915	Avezzano	FF	137	XI	VII
1873	S. Ginesio	FF	142	VIII	VI–VII

D: epicentral distance; I_0 epicentral intensity; I_S site intensity

observed in pervasively fractured zones where joints dip toward the slab, thus creating hanging rock faces. Failed blocks can have noticeable runout, endangering the slope below.

Large failures, up to thousands of cubic meters, can occur in both the pozzolana facies and the intensely fractured tuffaceous facies. The failed masses form fans consisting of blocks interspersed in a finer volcaniclastic mass (see Fig. 9).

Rock mass stability can be locally influenced by underground cavities. Vertical stresses in the underground pillars and tensile stresses on the vaults are comparable with uniaxial compressive and tensile strengths, respectively, of the pozzolana (see section 6.1). As a matter of fact several old quarry pillars are seriously damaged, especially at the slab base (Fig. 27), and many vaults of shallow or parietal chambers have collapsed, thus undermining the overlying structures or inducing cliff instability.

Cliff failures and landslides in the clayey slope are closely interrelated. An example is the progressive collapse of the S. Chiara rock spur related to the repeated landslides within the clayey slope below (*Cannicella* landslides). A landslide involved the clayey slope between April 7th and 8th 1900, during a series of rock slides which partly dismantled the overlooking rock spur (Fig. 9). Three landslides occurred between 1979 and 1984 in the *Cannicella* district during a series of failures and movements of the S. Chiara rock spur, as detected by a rod extensometer (Fig. 28). It is worth noting that the landslide scarps occur at the very foot of the cliff and that in 1979 the clay was exposed on the scarp face.

7.3 *Seismic hazard*

Despite only moderate seismicity of the area, archival sources report rock falls and severe building damage during historic earthquakes. Local seismic response analyses have been performed in the past to give quantitative support to the archival data (Muzzi et al., 2001, Lanzo et al., 2004).

Orvieto has suffered the effects of both near-field (NF) and far-field (FF) earthquakes (Tab. 2). The strongest NF event was the 1695 Bagnoregio earthquake, which produced the highest intensity at Orvieto and severe damage to buildings (e.g. fissures in the masonry walls of the Cathedral). Epicentres of FF events were in the Central Apennines (75 km < D < 140 km).

The strongest were: the 1328 and 1703 Norcia earthquakes, the 1751 Gualdo Tadino earthquake, the 1915 Avezzano earthquake, and the 1349 Aquila earthquake. Significant damage was produced only by the 1915 Avezzano earthquake (Cathedral, S. Giovenale church and several public buildings) and the 1349 Aquila earthquake ("many walls, large edifices, towers and palaces collapsed", according to medieval chronicles). Unexpectedly, buildings also suffered damage after weaker earthquakes, such as the 1873 S. Ginesio earthquake, which damaged the S. Domenico church. Rock falls were frequently recorded except for the farthest high-magnitude earthquakes.

One- and two-dimensional total stress analyses were performed. Two 2D symmetrical models were built along the maximum (1400 m) or the minimum (700 m) width of the slab (Fig. 29). Real accelerograms representative of the maximum historical NF and FF events were used as input motion.

The stratigraphy of the slab used in the analyses (Fig. 29) is representative of most, although not all, of the entire slab. From bottom to top it consists of a 30 m thick tuff layer, a 10 m thick layer of competent pozzolana and a 20 m thick soft pozzolana layer. The slab is superimposed on a 200 m thick clay substratum which overlies a sandstone-marly flysch formation (seismic bedrock).

The combination of various V_s and ρ values (Fig. 29) yields singular impedance ratios between materials. The clay-to-tuff impedance ratio, lower than 1.1, maintains the shear strain below the volumetric threshold (about 0.05% for clays of similar plasticity) and in turn prevents post-seismic deformations due to excess pore pressures in the clay deposit (Lanzo et al., 2004). Conversely, tuff-pozzolana and hard pozzolana—weak pozzolana impedance ratios (1.24 and 1.27, respectively) determine the main amplifying effects. 1D analyses (Fig. 30) indicate that the peak acceleration a_{max} reaches 0.22 g and 0.12 g at the slab top for NF and FF motions, respectively, with the sharper increase above the upper interfaces. The amplification ratio between a_{max} at the slab top and that at the bedrock is 1.8 (NF input) and 2.8 (FF input), richer in frequencies close to the resonance frequency of the entire sequence overlying the bedrock (0.6 Hz).

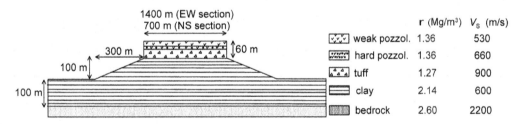

Figure 29.   Sketch of the 2D numerical model used for the seismic response analyses.

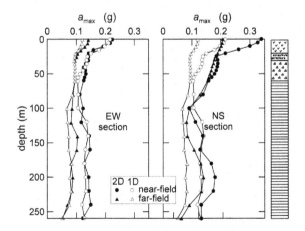

Figure 30.   Vertical profiles of peak acceleration at the hill centre, computed through 1D and 2D analyses (from Lanzo et al., 2004, modified).

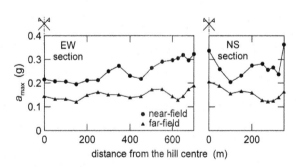

Figure 31.   Peak acceleration along the slab top computed through 2D analyses (from (Muzzi et al.,) 2001, modified).

Topographic effects produce further amplification. 2D analyses along the wider E-W section show that a_{max} at the centre of the slab is similar to that calculated through 1D analyses, but it increases significantly towards the edge, up to 0.32 g and 0.19 g for NF and FF input motion (Fig. 31). Topography also influences a_{max} far behind the cliff edge, especially along the shorter N-S model (Fig. 31): a_{max} at the centre of the N-S section is 0.34 g (NF) and 0.21 g (FF). Over the whole slab top the acceleration peak is associated with harmonics having a period in the range 1.5–2.0 s. This amplification, due to a 2D normal mode of oscillation of the slab, adds to that generated by the 1D resonance period (1.7 s) and sharpens the sensitivity of tall buildings to low-frequency-reach earthquakes, as documented by historical reports.

8   THE COMPREHENSIVE INSTABILITY MECHANISM

A more complete understanding of the complex instability mechanisms requires a comprehensive analysis which considers the interaction between the clay slope and the rock cliff. Cliff failures and deformations represent the response of the pyroclastic slab to the erosion of the clay formation, which at present proceeds through continuous slope movements and occasional landslide events.

The bulk effect of the erosional process that created the Orvieto hill is well depicted by the stress-strain finite element (FE) and finite difference (FD) analyses conducted by Cecere & Lembo-Fazio (1986) and Verrucci (1998). Simplified 2D models of the hill based on the geotechnical characterization

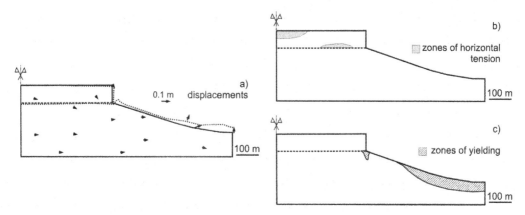

Figure 32.   FEM analyses. Displacements and deformed boundaries a), zones in tension within the slab b), zone at yielding within the clay slope c) (modified from Cecere & Lembo-Fazio 1986).

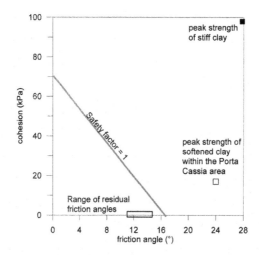

Figure 33.   Comparison of the strength mobilized in the 1900 Porta Cassia from 2D limit equilibrium analyses and shear strength from laboratory tests under peak and residual conditions.

described in section 6 were set up. The stress history was subdivided into four stages: i) initial deposition of the clay deposit; ii) erosion of its upper 150 m; iii) deposition of the volcanic materials; iv) erosion to the present day morphology.

The adoption of a strain-softening constitutive model for the clay material complies with some of the structural features and failure mechanisms observed at the slab margin (Fig. 32).

Another effect of the erosional process is the formation of zones with high plastic shear strains beneath the slab margin and along the slope, which can be linked to the widespread softening of the uppermost part of the clay formation (as indicated by geotechnical tests). At the outcrop scale evidence of this phenomenon is provided by systematic vertical joints in the softened part of the clay formation. If softening is considered, tensile stresses at the upper part of the cliff increase, thus supporting upward fracture propagation.

Deep movements, recorded by monitoring instrumentation, extend across the entire slope along slip surfaces which fully develop within the softened portion of the clay formation. Strength mobilized along these slip surfaces is close to residual values (Fig. 33), thus indicating that slope movements utilize pre-existing surfaces produced by ancient failures that are no longer possible under the present morphological and hydraulic conditions.

The marginal fractured portion of the slab are deprived of lateral support by these continuous deep slope movements, which also remove the talus debris mantling the cliff foot that could otherwise provide local confinement. Thick rock slices or large spurs can slide, topple, or collapse, especially if the rock mass is intensely fractured or where the pozzolana facies prevails.

Earthquakes have an additional but not fundamental role on the stability of the slab margin. Accelerations induced by amplification effects at the cliff edge do not influence the stability of large portions of the slab margin but can trigger falls of marginally stable blocks on the cliff face, according to historical reports.

9 THE MODERN PROTECTION/PRESERVATION PROGRAMME

Different forms of remediation have been undertaken over the centuries to stabilise the cliff, but it has only been since the end of the past century that a more holistic approach has been developed, which was devoted to long-term protection and preservation rather than just reacting to immediate emergencies.

In 1978, 41 years after the Italian state had taken charge of the protection of Orvieto (see section 5.2), a wide-ranging and multidisciplinary protection/preservation programme was instigated with the passing of Parliamentary law n. 230 of 25/05/1978 ("Urgent actions for the consolidation of the cliff of

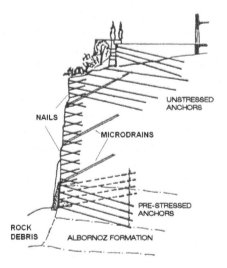

Figure 34. Schematic section of the consolidation works carried out on the cliff (modified from Cestelli-Guidi et al., 1983).

Figure 35. Reconstruction of the wall under the Albornoz fortress (after Soccodato et al., 2013).

Figure 36.   Stabilization of the clay slope beneath the S.Chiara Nunnery. The arch-shaped retaining wall (lower-left part) and the shaft from which tubular drains depart are visible (from Soccodato et al., 2013).

Orvieto and the hill of Todi to preserve the landscape and historical, archaeological and artistic heritage of the two towns"). The programme was completed thanks to a further legislative act of the Italian Parliament (n. 545 of 29/12/1987) which allocated funds for "Provisions for the final consolidation of the cliff of Orvieto and the hill of Todi".

This change in attitude can likely be attributed to a broadminded public-works policy based on the realisation that: i) such widespread and differentiated stability problems could not be faced with ordinary regional budgets; and ii) the close interaction between the complex instability phenomena and an ancient and populated town required a comprehensive intervention.

Two criteria guided the design of the original protection and preservation programme (Cestelli-Guidi et al., 1983): a) remove the disturbances induced by human activities; and b) restore the overall static stress conditions that developed during the geological evolution of the rock slab.

Initially, remedial measures (falling within design criteria (b)) urgently addressed the stabilization of selected areas of the cliff, where failures and large movements had recently occurred which posed a serious threat to human safety and building and infrastructure security. The cliff was reinforced using short passive bolts (nails) and moderately pre-stressed or passive deep anchors (Fig. 34), and sub-horizontal drains were installed to prevent an increase in water pressures along the joints that extend behind the tuff blocks. This scheme has been successively replicated in many other parts of the cliff (Soccodato et al., 2013), with adaptations for specific local geotechnical conditions.

Subsequently, issues related to design criteria (a) were addressed via the following main countermeasures:

– reduction of infiltration within the urban areas by renewing street pavements and restoring the sewer and aqueduct networks to eliminate leaks inferred by chemical monitoring of the spring waters seeping at the cliff foot (Conversini et al., 1977);
– collection into the pre-existing gullies of waste water from the slab and of surface water over the slope (through a drainage delivery system) to limit infiltration and erosion;
– regulation of gully hydraulics using check dams;
– stabilization and protection against erosion of the gully banks by remodelling and replacement of the clay debris with a thick layer of compacted volcaniclastic material reinforced with geogrids (Coluzzi et al., 1995).

In addition to the reinforcement scheme reported in Figure 34, the remedial works related to design criteria (b) can be summarized as follows. Restoration of historic retaining/lining walls (e.g., the wall underneath the Albornoz Fortress shown in Figure 35), including substitution of the more degraded portions and static rehabilitation by underpinning and reinforcement with nailing and injections.

Stabilization works on the clay slope involved not only the gully banks, as mentioned above, but also areas far from the gullies.

First the north-eastern part of the slope, at the eastern margin of the 1900 Porta Cassia slide area, was involved in continuous slope movements down to a depth of 14 m with an average displacement rate of 60 mm/year (Pane and Martini, 1996). This section was stabilised with a deep drainage system that consisted of fans of sub-horizontal tubular drains drilled from a series of hydraulically connected shafts.

Second the *Cannicella* zone, near the steep slope underneath the S. Chiara nunnery, had been affected by two landslides (1983 and 1984) that involved the talus and underlying clayey debris soon after the cliff strengthening reported in Figure 34. These events, which occurred in a particularly degraded area (see section 7), highlighted the need to accelerate stabilization of the clay slope. This was actuated through drainage wells and an imposing, arch-shaped, anchored retaining wall. The wall, buried within the debris to maintain an unaltered landscape, has the function of confining the thick debris cover and the base of the cliff (Fig. 36).

An extensive monitoring system was installed in the early 1980s to monitor the hill behaviour and to evaluate the effectiveness of the adopted remedial measures. The system, set up by the Regional Government of Umbria in coordination with the Municipality of Orvieto, and managed by the "Observatory for permanent control and maintenance", consists of a topographic network and more than 200 instruments, including piezometers, inclinometers and extensometers in the cliff (Soccodato et al., 2013).

10 FINAL REMARKS

The geological and geotechnical conditions of the site have intimately influenced the history of Orvieto since its foundation. The relationship between the town and its "subsoil" is intrinsically complex due to the mechanisms controlling the stability of the hill, but it has been further complicated by different human activities, each with a variable extent in time.

As a result of the protection-preservation programme started in the early 80's, the human impact on factors affecting the instability mechanisms has been minimized, the stability of the rock mass and clay slope in critical areas has been restored, and the rock mass all around the rock slab has been strengthened. The result is that the frequency of failures has been drastically reduced.

However continuous effort is needed to maintain the efficiency of these successful remedial and prevention measures.

Monitoring data, released by the Observatory for permanent control and maintenance and by the independent monitoring system of the National Research Council, indicate that the clay slope, in various areas of the hill, is experiencing very slow shallow and deep movements. These deformations are the "engine" of instability mechanisms, in that they induce unavoidable deformations at the slab margin, which have the significant potential to damage buildings and infrastructure.

This ongoing process implies that a second major continuous-monitoring effort is needed to timely recognize changes in deformation trends in the medium- and long-term, thus preventing the development of failure or large displacements at the slab margin.

Prevention policy could be aided by coupling monitoring data to refined geotechnical models in order to formulate instability scenarios at selected areas characterized by different controlling mechanisms. Monitoring data should be integrated to provide missing information and site specific geological and geotechnical investigations should be conducted to account for local stratigraphy and hydraulic conditions, which vary widely from one zone to another. Site specific investigations should also be conducted to better understand the static conditions of underground cavities and the local site response to seismic actions, which, in spite of the low seismicity of the area, have been a source of damage to monumental edifices during the past centuries. Finally long term scenarios should also account for climate changes which could modify the clay slope evolution and the related response of the slab margin.

ACKNOWLEDGEMENTS

The research would not have been possible without the contribution of the M.Sc. Students, who carried out their thesis on topics related to the stability of the Orvieto Hill. Those who are not explicitly cited in the text are: N. Buttafoco, F. Coni, D. D'Alberti, F. Lucci, E. Moscatelli and L. Rosa.

Monitoring measurements and many of the laboratory tests were conducted by the Technicians of CNR-IGAG: R. D'Inverno, A. Cittadini, E. Tempesta, P. Millozzi and M. Paciucci.

REFERENCES

Cecere, V. & Lembo Fazio, A. 1986. Condizioni di sollecitazione indotte dalla presenza di una placca lapidea su un substrato deformabile. *Proc. XVI Convegno Nazionale di Geotecnica, Bologna* 1: 191–202.

Cestelli-Guidi, C., Croci, G. & Ventura, P. 1983. The stability of the Orvieto rock. *Proc. 5th ISRM Congress, Melbourne*, C31–C38.

Chandler, R.J. 1984. Delayed failure and observed strength of first time slides in stiff clays: a review. *Proc. 4th Int. Symp. on Landslides, Toronto*, University of Toronto Press, 2: 19–25.

Coluzzi, E., De Carolis, P., Rimoldi, P. & Soccodato, C. 1995. Rinforzo mediante geogriglie delle pareti del Fosso della Civetta alla base della Rupe di Orvieto. XIX *Convegno Nazionale di Geotecnica, Pavia*, 1: 201–210.

Conversini, P., Lupi, S., Martini, E., Pialli, G. & Sabatini, P. 1977. Rupe d'Orvieto: indagini geologico tecniche. *Quaderni della Regione Umbria—Supplemento*, p. 82.

De Marchi, L. 1897. Provvedimenti di sicurezza nelle cave di tufo e pozzolana esistenti al perimetro della città di Orvieto a tutela dell'abitato. *Rivista del Servizio Minerario*: 277–282.

Ferrovie Meridionali 1888. Constatazione delle linee componenti la rete adriatica, Linea Roma—Orte—Chiusi., Roma. *Technical Report*, Rome.

Lanzo, G., Olivares, L., Silvestri, F. & Tommasi, P. 2004. Seismic response analysis of historical towns rising on rock slabs overlying a clayey substratum. *Proc. 5th Int. Conf. on Case Histories in Geotechnical Engineering, New York, 13–17 April 2004*.

Lembo-Fazio, A., Manfredini, M., Ribacchi, R. & Sciotti, M. 1984. Slope failure and cliff instability in the Orvieto hill. *Proc. 4th Int. Symp. on Landslides, Toronto, 16–21 September 1984*, Toronto: Canadian Geotechnical Society, 2: 115–120.

Manfredini, M., Martinetti, S., Ribacchi, R. & Sciotti, M. 1980. Problemi di stabilità della rupe di Orvieto. *Proc. XIV Convegno Nazionale di Geotecnica, Firenze, 28–31 October 1980*, Rome: Associazione Geotecnica Italiana, 1: 231–246.

Martini, E. & Margottini, C. 2000. Le frane storiche di Todi e Orvieto. CD-ROM. Osservatorio della Rupe di Orvieto e del Colle di Todi.

Muzzi, M., Lanzo, G., Tommasi, P. & Ribacchi, R. 2001. Analisi numerica della risposta sismica del colle di Orvieto. *CD ROM of Proc. of X Italian Congress ANIDIS, Potenza*.

Pane, V. & Martini, E. 1996. The preservation of historical towns of Umbria: The Orvieto case and its observatory. In Carlo Viggiani (ed.), *Geotechnical Engineering for the Preservation of Monuments and Historic Sites; Proc. of the Arrigo Croce Memorial Symposium, Napoli, 3–4 October 1996*. Rotterdam: Balkema. 489–498.

Perali, P. 1919. *Orvieto, note storiche di topografia e d'arte dalle origini al 1800*. Orvieto: Marsili.

Soccodato, F.M., Tortoioli, L., Martini, E. & Mazzi, M.A. 2013. The Orvieto Observatory 2013. In Margottini C, Canuti P, Sassa K (eds), *Landslide Science and Practice* 6: 547–560, Berlin: Springer.

Stiny, J. 1930. Die Bausteine Orvietos und ihre Verwitterung. *Geologie und Bauwesen* 3: 1–40.

Tommasi, P., Boldini, D., Caldarini, G. & Coli, N. 2013. Influence of infiltration on the periodic re-activation of slow movements in an overconsolidated clay slope. *Can. Geotech. J.* 50: 54–67. dx.doi.org/10.1139/cgj-2012-0121.

Tommasi, P., Pellegrini, P., Boldini, D. & Ribacchi, R. 2006a. Influence of rainfall regime on hydraulic conditions and movement rates in the overconsolidated clayey slope of the Orvieto hill (central Italy). *Can. Geotech. J.* 43: 70–86. doi: 10.1139/T05–081.

Tommasi, P., Rotonda, T. & Ribacchi, R. 2006b. Caratterizzazione geotecnica della pozzolana di Orvieto. In G. Urciuoli (ed.), *Questioni di ingegneria geotecnica, Scritti in onore di Arturo Pellegrino*, Hevelius: Benevento, 677–700.

Tommasi, P. & Ribacchi, R. 1998. Mechanical behaviour of the Orvieto tuff. In A. Evangelista & L. Picarelli (eds), *The Geotechnics of Hard Soils—Soft Rocks; 2nd Int. Symp., Napoli*, Rotterdam: Balkema, 2: 901–909.

Tommasi, P., Ribacchi, R. & Sciotti, M. 1986. Analisi storica dei dissesti e degli interventi sulla Rupe di Orvieto: un ausilio allo studio dell'evoluzione del centro abitato. *Geologia Applicata e Idrogeologia* 21: 99–153.

Tommasi, P., Ribacchi, R. & Sciotti, M. 1997. Geotechnical aspects in the preservation of the historical town of Orvieto. In Carlo Viggiani (ed.), *Geotechnical Engineering for the Preservation of Monuments and Historic Sites; Proc. of the Arrigo Croce Memorial Symposium, Napoli, 3–4 October 1996*. Rotterdam: Balkema. 849–858.

Verrucci, L. 1998. Caratterizzazione meccanica del tufo di Orvieto e stabilità della rupe. M.Sc. Thesis, Università di Roma "La Sapienza", Roma.

Vinassa de Regny, P. 1904. Le frane di Orvieto. *Giornale di Geologia Pratica* 1: 110–130.

Geotechnics and Heritage – Bilotta, Flora, Lirer & Viggiani (eds)
© *2013 Taylor & Francis Group, London, ISBN 978-1-138-00054-4*

A paradise inhabited by devils? The geotechnical risks in the city of Napoli and their mitigation

A. Evangelista & C. Viggiani
University of Napoli Federico II, Napoli, Italy

ABSTRACT: Napoli is probably the only large European city whose historical center is still inhabited by people, and not only by offices, banks and shopping malls; it has been declared World Heritage by the UNESCO. It is affected by quite a number of problems, among whom some significant geotechnical risks. The mitigation of such risks in the city has to be considered hence as a contribution to the safeguard of an outstanding cultural heritage, besides contributing to a better future for the Neapolitans. The paper reports the main geotechnical problems affecting the city and the initiatives undertaken to solve them. Some interesting cases are discussed in detail.

1   INTRODUCTION

An ancient image, dating back to the Middle Ages, spread in Europe in the XVIII Century by people traveling through the *Grand Tour* and resumed later by Benedetto Croce (Galasso, 2006), portrayed Southern Italy as a paradise inhabited by devils. It was meant that, in spite of the mild climate and fertile soil, the peoples were poor for their fault, simply because they were indolent and dissolute. This metaphor is certainly questionable if applied to Southern Italy, but it applies well to the subsoil of Napoli, since the engineering properties of the soils are on the whole rather favorable and most of the serious geotechnical problems affecting the urban area are essentially a product of the inexorable activity of the men; as a matter of fact, the oldest human settlement in the area dates back to almost three millennia.

In the last decades two relatively new theses have been elaborated in the realm of heritage preservation: (i) a historical center is a physical system to be protected and restored as a whole, and (ii) it has a peculiar nature: not a lifeless object to be visited and admired as the artworks in a museum or a traditional monument, but an inhabited set, where man may reconcile with his environment.

The historical center of the city of Napoli is a heap of cultural goods, monuments, pictures, sculptures, available for a public fruition; it has been declared World Heritage by the UNESCO. It is probably the only historical center of a large European city still inhabited by people, and not only by offices, banks and shopping malls. It is affected by quite a number of problems of economic, social and physical nature; among the latters, a number of geotechnical risks. Any action aimed at the mitigation of geotechnical risks in the city or at making its fruition easier has to be considered as a contribution to the safeguard of an outstanding cultural heritage. Furthermore, such actions contribute to pursuing the utopia of a better future for the poor devils inhabiting Napoli.

2   GEOLOGICAL FRAMEWORK

The city is located at the southern border of an upland plain of volcanic origin known as *Terra di Lavoro*, formed around 35,000 years ago as the results of a gigantic eruptive explosion whose main product is the grey Neapolitan tuff, which forms the base of the upland plain. After the grey tuff eruption, there was a long period of volcanic inactivity followed by a number of smaller eruptions between 13,000 and 11,000 years bp, from volcanoes located west of the present city, in the region of the so called Phlegrean Fields. The main material produced by these eruptions is the pyroclastic material known as *pozzolana*, that is volcanic ash made by glassy vesicular particles with interposed pumices, with a grey to yellowish color and deposited essentially with mechanisms of pyroclastic flow. It is a non-plastic, more or less silty sand.

Under some conditions of temperature and chemical composition, after the deposition the pozzolana develops cementation bonds, transforming into the Neapolitan Yellow Tuff. In general, the tuff is found in the area closer to the eruptive center while moving away the cohesionless pyroclastic materials prevail. In fig. 1 the boundaries of the major Phlegrean calderas (grey tuff and yellow tuff) are reported.

The so called third period of activity of the Phlegrean volcanoes, around 10,000 years bp, gave origin to deposition of pyroclastic materials essentially by fall.

At the end of each eruption phase generally the central part of the caldera collapses and sinks, destroying part of the ring of material accumulated around the volcanic mouth, and leaving a structure in the form of an arc of circle.

The remains of the volcanic structures superimpose one over the previous one, concealing the latter and resulting in a territory characterized by an uneven morphology (fig. 2): narrow coastal plains bounded byrelatively steep slopes, reaching over 400 m above sea level.

The pyroclastic soils deposited over the slopes are easily eroded by rain water, and hence deep gorges have been excavated in the slopes.

The backbone of the area is constituted by volcanic rocks, covered by pyroclastic soils (pozzolana, volcanic ashes, pumices) in primary deposition on the hills and eroded, transported and re-deposited as alluvial sediments in the coastal plains. Within the depth of engineering interest the most widespread rock is the Neapolitan yellow tuff, a soft rock with a porosity as high as 30% to 50% and a compressive strength ranging from 3 to 10 MPa. It has been deposited above the pre-existing hills covering them with a thickness ranging from 30 to 180 m. The pyroclastic soils covering in turn the yellow tuff are essentially cohesionless more or less silty fine sands with a very low unit weight in the range between 14 and 16 kN/m³. They are characterized by a friction angle of the order of 35° and a small and very variable cohesion, in the range between 0 and 20 kN/m², partly due to fragile cementation bonds and partly to suction.

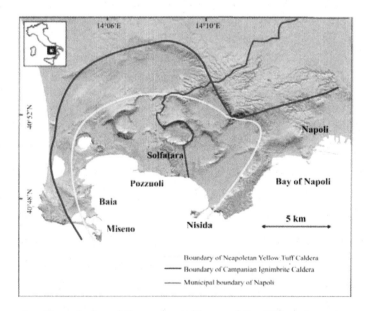

Figure 1.   Boundaries of the major calderas of the Phlegrean volcanoes.

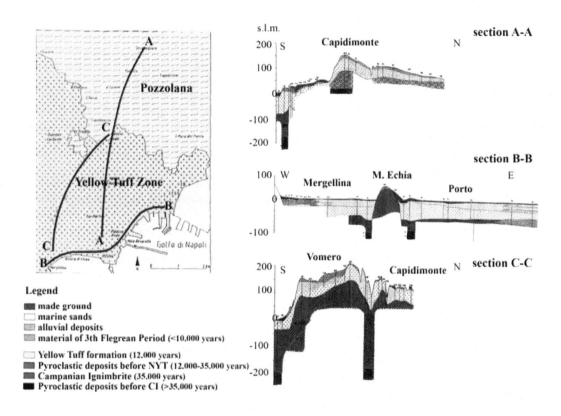

Figure 2.   Geological sections (De Riso, Bellucci, 1998).

The water table is found a few meters above sea level, being thus rather shallow near the coast and progressively deeper below the hills; the pyroclastic soils are hence mostly unsaturated. They exhibit a metastable structure subjected to collapse with significant volume decrease at soaking with water.

Fig. 2 reports some schematic geological sections of the urban territory.

3   GEOTECHNICAL RISKS

The main geotechnical problems affecting the urban area are the following.

- Fast flow slides are triggered by heavy rainfall in the thin pyroclastic cover of the steep slopes (fig. 3). The slide volume is generally rather small (some tens to some hundreds cubic meters); during last decades, however, an increasing number of buildings have been located at the toe of the slopes, and their inhabitants are thus exposed to a serious landslide risk.
- The tuff outcrops in natural steep cliffs and vertical man-made cuts, reaching the height of many tens of meters and a total surface of about 4 km^2 in the urban area. They are affected by different instability phenomena connected to surface weathering and the occurrence of discontinuities, and threaten the underlying built environment by risk of rock falls (fig. 4).
- In the urban area there are some 250 km of retaining walls higher than 3 m. The large majority of themare ancient structures in tuff masonry (fig. 5), whose characteristics are not known and whose stability is rather uncertain; frequent collapses occur during intense rainfall events.

An intricate network of cavities has been excavated within the tuff formation over the centuries for different purposes. Most of them have been subsequently abandoned, and in many cases their characteristics or even their existence are unknown. At the time being there are 733 known cavities and only 2/3 of them have been surveyed. They raise a variety of problems, including collapses and sinkholes threatening the overlying buildings and infrastructures.

Figure 3. Flow slide in the pyroclastic cover of a slope.

Figure 4. Piazza S. Luigi. The vertical walls of an ancient tuff quarry.

Figure 5. Typical retaining walls: (top left) wall lined with lava stones in Via Caravaggio; (top right) wall with buttresses in Via Bonito; (bottom left) wall with a release surface for contact with the backfill in Petraio and (bottom right) wall to be consolidated in Via Aniello Falcone.

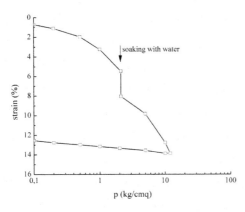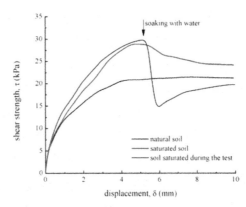

Figure 6.   Volume change and shear strength reduction of unsaturated pyroclastic soil at soaking with water.

- The sewerage and drainage network in the central area dates back essentially to 1910; the main sewers have a total length over 600 km (Rasulo, 2000). Due to the expansion of the city following the World War II, many ducts are at present inadequate and undergo frequent ruptures during intense and prolonged rainfall events. Collapse of ducts and consequent sinkholes occur frequently all over the city.
- Furthermore, the unsaturated pyroclastic soils are very sensitive to the effects of water; they are easily eroded by rain water running over the hill slopes or seeping toward underground openings, and experience a structural collapse at soaking with water with volume reduction, and hence settlement, and shear strength reduction (fig. 6). This is at the origin of a number of damages to the buildings as a result of water leaks from the aqueduct and sewage network.

4   PREVIOUS INVESTIGATIONS AND ATTEMPTS OF MITIGATION MEASURES

At the end of XIX century an engineer of the Municipality, G. Melisurgo, first realized the need for a systematic study of the subsoil of the city and the intricate network of tunnels, canals, wells. He reported the results of his own investigations in an interesting and informative book (Melisurgo, 1889).

The frequency of damages and troubles connected to the subsoil was continuously increasing after the World War II, and prevention and mitigation measures were strongly required by public opinion. A first Study Commission on the subsoil of the city was thus installed by the Municipality in 1966, and produced a broad final report (Comune di Napoli, 1967).

The Commission concluded that troubles and damages were essentially a consequence of actions of the men, rather than being connected to the characteristics of the subsoil. In the historical centre of the city small damages had occurred for a long time, essentially because of small water infiltrations in the subsoil and for the poor quality and age of the buildings, often continuously tampered with new openings and addition of further floors. The water infiltrations, and hence the troubles, increased in frequency and gravity with the construction of the new aqueduct at the end of the XIX century. In the areas of recent edification on the hill slopes, small damages were rare because of better quality and younger age of the buildings; large damages, on the contrary, frequently occurred in connection with a chronic shortage or inadequacy of infrastructures following a rapid and chaotic expansion.

The Commission suggested some measures, as new regulations to be included in the city building code, the institution within the Municipality of a technical service dealing with the subsoil, the execution of site investigations and the redaction of subsoil maps; some of these suggestions were actually accepted and partly actuated.Unfortunately, these measures revealed inadequate. In 1967/68 the announcement of a new national town planning law, to be applied in one year time, produced the fear of substantial limitations of the building activity and consequently an abrupt increase of residential constructions with small or no attention to the infrastructures and the subsoil. To face the consequent increase of troubles, the city government installed a new Commission. In its final report (Comune di Napoli, 1972), the conclusions of the 1967 Commission were repeated and detailed.

Following the widespread damages to the built environment produced by the 1980 Irpinia—Campania earthquake, new Commissions have been installed both by the Municipality and by the national Government; a significant amount of resources was made available by the Government and a number of interventions were actually carried out, especially to improve the sewage and drainage network. The outcome of these works, however, has been rather disappointing, essentially for the lack of coordination; many of the new works have never been put in operation.

5 THE SUBSOIL EMERGENCY COMMITTEE, 1999–2007

1996 has been a bad year for Napoli; it began in January with the collapse of a road tunnel under construction at Secondigliano, in the Northern outskirts of the city, with 9 casualties (fig. 7), and ended in December with a sinkhole in Miano, near the Royal Garden of Capodimonte, killing two persons. Following these occurrences, the Italian Government appointed the Mayor of Napoli as a delegate for the Subsoil Emergency, with the task of conceiving the necessary prevention and remedial measures and initiating the design and construction activity.

The "emergency" was actually as long as 9 years. The law allowed the Delegate Mayor to install a Scientific Committee; this was actually done, designating a group of geologists and engineers coming from the cadres of the Municipality and from the Napoli Universities.[1]

A broad and careful scrutiny of the available knowledge led the Committee to realize that the geotechnical problems of the city have some common features. They are widespread over the urban area, with somewhat repetitive characters; each category of problems can thus be studied as such, and not as a mere accumulation of single cases. They cause high risk situations, requiring often immediate countermeasures. A huge amount of resources is needed to significantly mitigate the risk; accordingly, for economical and management reasons, some prioritization of the interventions is mandatory.

The activity of the Committee moved along three directions: (i) management of the emergency, with immediate interventions aimed at solving unbearable situations; (ii) investigations aimed to collect the data needed to define the long term interventions and (iii) preparation of a plan of the long term interventions, including an order of priority.

Figure 7. Collapse of a road tunnel under construction at Secondigliano. The collapse involved a gas pipe, producing a spectacular fire.

---

1. A. Evangelista has coordinated the investigations carried out on behalf of the Committee by the CUGRi, Research Centre of the University of Napoli Federico II; C. Viggiani has been a component of the Committee.

For each of the classes of geotechnical problems mentioned before (fast flow slides, rock falls, retaining structures, underground cavities and drainage and sewerage network), the following steps have been followed:

– inventory of the existing documentation and planning of integrative investigation, if needed;
– filing of all the information in an informatic archive, allowing easy retrieval and cross referencing;
– detailed examination of a small number of typical cases; design and implementation of the remedial measures;
– formulation of guidelines for further generalized interventions;

The activity of the Committee has been by far too broad to be synthesized in a paper; in the following paragraphs some information is given on some of the items above.

6 UNDERGROUND CAVITIES

6.1 *The problem*

As reported above, an intricate network of cavities has been excavated within the yellow tuff formation over the centuries for storage, aqueducts, roads and railways, water reservoirs, underground quarries, military purposes.

The two principal and widespread systems of cavities are those of the ancient aqueducts and the underground quarries for the extraction of the tuff as a construction material. The aqueduct of the Bolla, or Volla, bringing the water drained 10 km apart in the plain north east of the city, entered the urban territory at an elevation of 13 m above sea level and hence it could serve only the lower part of the town. It dates back to Greek and Roman times, but has been continuously modified and enlarged during the centuries. The Claudius aqueduct, constructed in the I century BC, came from springs in Serino, 100 km apart, and was conceived to feed the villas and harbor on the coast west of Napoli; a lateral branch was however serving the city. The aqueduct of Carmignano, built in the XVII century and bringing the water from a river 55 km away, allowed an increase both of the available amount of water and of the area served. Within the city, these aqueducts consisted in a network of small tunnels bringing the water to a myriad of subterranean tanks (fig. 8; 9).

People drew water from the tanks through wells; each tank was served by one or more wells. It is evaluated that in the XIX century the number of tanks was as high 4000 and the number of wells probably much higher; only about 2200 of these wells have been rediscovered. The ancient free surface aqueducts have been in use till the end of XIX century, with serious hygienic risks; only in 1885, after a grave cholera epidemic, a modern aqueduct with pressure pipes has been realized.

The underground tuff quarries have a size much larger than the water cisterns (fig. 10). The most ancient are located within the city walls, as for instance the Palamoniae Caves, devoted to the cult of god Mitra (fig. 11). Later on, the large underground quarries were located at the outer boundaries of the city center, at the toe of the surrounding hills; they were entered through the erosion gorges, where the tuff outcropped, or by means of vertical wells. Cavities with a volume of over 100,000 m^3 are not uncommon.

The road or railway tunnels are also widespread, starting from the Roman tunnels crossing the Posillipo ridge (Grotta di Seiano, Crypta Neapolitana) and ending to the present metro lines still under construction.

All these cavities raise quite a number of problems: collapse of the roof, fall of blocks inside the cavity, sinkholes produced by internal erosion of the cohesionless covering the tuff with transport of fines to the cavities.

The first stage of the activity of the Committee has been the inventory of the existing cavities and the collection of all the available information such as the main geometrical data (surface, volume, maximum and average height, number of pillars, maximum span of the cover, shape of the section), the data needed for a first evaluation of the statical conditions (thickness of the tuff cover, thickness

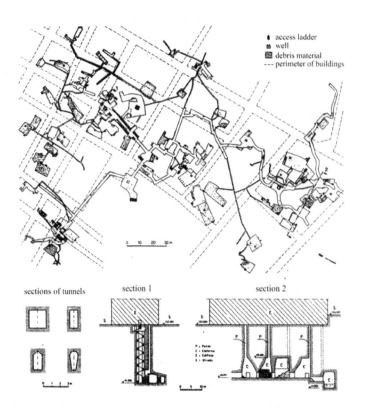

Figure 8.   The network of tanks and small tunnels of the ancient aqueduct in a zone of the city centre.

Figure 9.   A typical aqueduct tank.

of soil above the tuff, occurrence of fractures and block fall), the occurrence and type of building and infrastructures at the surface above the cavity, to evaluate the vulnerability.

The existence of 733 large cavities is documented; for 560 among them a topographical survey is available, while the remaining 173 are listed, but not surveyed. Only in 171 cases the information were complete enough to allow the analyses that will be reported later. In 221 cases it is evident that

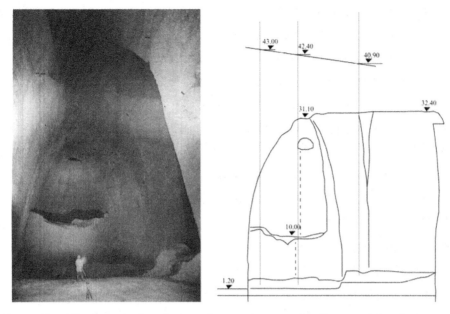

Figure 10. The large cavity of an ancient underground tuff quarry; San Rocco cavity.

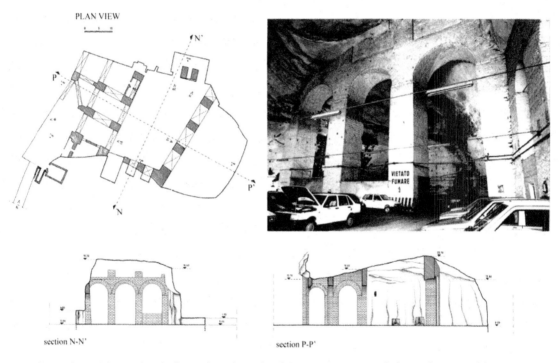

Figure 11. The cavity dedicated to the cult of Mytra; at present it is used as a parking area.

the existing survey cover only a part of a larger and unknown cavity. There are a number of signs that the actual number of cavities may be much larger; the search for new cavities and the completion of survey will hence continue in the future. At the time being, the total volume of known cavities is about 6 millions of cubic meters; two thirds of the cavities are smaller than 5,000 m^3, and one third smaller than 1,000 m^3. The thickness of the tuff cover over the cavity is larger than 5 m only in the 27% of the 184 cases in which the figure is available, while is less than 2 m for 35% of the cases.

83

6.2 *Stability analysis*

A complete analysis of the state of stress around a cavity should ideally take into account the three dimensional shape, the existence and orientation of the discontinuities, the original in situ stress and its modifications following the excavation, the properties of the materials. Such data are not available in the large majority of cases, and however the number of cavities to be analysed hinders a detailed examination. At least in a first stage, a simple analysis has been preferred that could be of guidance for establishing a priority level for further investigations and interventions.

The expeditious analysis carried out considers separately the state of stress in the pillars and in the roof. The stress in the pillars is evaluated by simply considering the load acting on the influence area of the pillar (fig. 12); a nominal factor of safety may be obtained as the ratio between the uniaxial compressive strength σ_c of the tuff and the stress in the pillar σ_{pil}. In the framework of a simple expeditious evaluation, the compressive strength of the tuff has been assumed equal to 3 MPa, and factors as the occurrence of fractures in the pillar or the influence of time and core size (Evangelista *et al.*, 1980) have been not been taken into account.

The safety of the roof has been evaluated referring to a local and a general failure mechanism (fig. 13). In the former the collapse consists in the detachment and fall of a block; in the latter the whole roof collapses. A parametric analysis has been carried out in the hypothesis of plane strain by numerical computations and by the simplified procedure reported in fig. 14. Based on the available experimental evidence, the tensile strength of the tuff has been assumed equal to 1/10 of the compressive strength. The results obtained are presented in fig. 15. The diagrams in fig. 15 may be used to evaluate the mini-

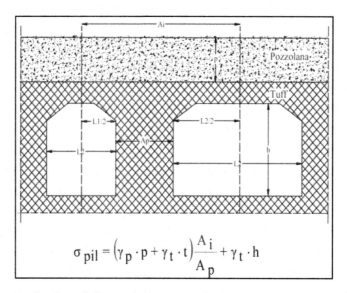

$$\sigma_{pil} = \left(\gamma_p \cdot p + \gamma_t \cdot t\right)\frac{A_i}{A_p} + \gamma_t \cdot h$$

Figure 12. Evaluation of the vertical stress at the base of a pillar by influence area.

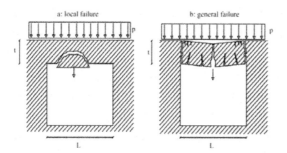

Figure 13. Failure mechanism of the cavity roof: a) local failure; b) general failure.

84

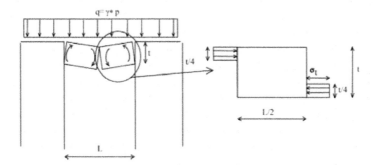

Figure 14. Simplified mechanism of general failure of the roof.

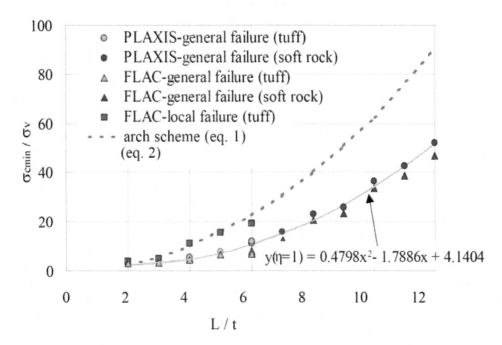

Figure 15. Results of the stability analyses of the cavity roof.

Table 1. Evaluation of the safety of cavities.

σ_{pil}, σ_r (MPa)	Factor of safety*	Stress level	Safety
<1	>3	Low	High
1 to 2	3 to 1.5	High	Low
>2	<1.5	Very high	Insufficient

*The factor of safety is the lowest between the pillars and the roof; local failure has been considered for the roof.

mum value $\sigma_{c,min}$ of the compressive strength needed to have a factor of safety equal to unity against both local and global failure, for a given set of geometrical conditions (span L, thickness of the roof t, depth below the ground surface and hence overburden pressure).

Being the analyses carried out rather rough, the results have been used in a semi-qualitative way defining three classes of stress level and safety, listed in Table 1.

About 30% of the analysed cavities revealed a low or insufficient safety level. As it was to be expected, the larger cavities exhibit in general the highest stress level; in terms of surface, hence, the cavities with low or insufficient safety are almost 60% of the total.

6.3 *Remedial measures*

There are various possible intervention techniques to stabilize the cavities, and the most suited one has to be selected depending on the type of cavity and the technical problem to solve. The most common technical problems are: (i) stabilisation to make the cavity available for fruition; (ii) intervention aimed only to stabilise the ground surface and the building and infrastructures existing above the cavity; (iii) stabilisation of the shafts in pozzolana; (iv) stabilisation of small tunnels within the pozzolana.

Among the possible intervention techniques we can list: (i) total or partial filling of the cavity with a suitable material; (ii) consolidation and retaining structure at the interior of the cavity; (iii) deep foundations for the building above, crossing the cavity and bearing below its bottom; (iv) lining of the shafts; (v) improvement of the cohesionless soils.

Some indications are reported in Table 2.

A significant example of filling of a large cavity is that of Cupa Spinelli (fig. 16, 17), an abandoned underground tuff quarry with a surface of about 4500 m^2 and an average height of 20 m.

Table 2. Intervention techniques.

Type of cavity	Statical conditions	Possible recovery	Interventions
Abandoned large underground quarries	Generally poor	Very expensive; uncertain usefulness	Total or partial filling from the surface
Large underground quarries still under use (cult)	Generally satisfactory	Finalised to public fruition	Only local, generally at the roof
Tunnels and tanks of the ancient aqueducts	Local problems possible in the shafts and in some tanks	Reconstruction of some stretch of the historical network	New connections, ventilation, lifts
Small cavities without historical interest	Generally good, except some shafts	Useless	Filling without further investigations

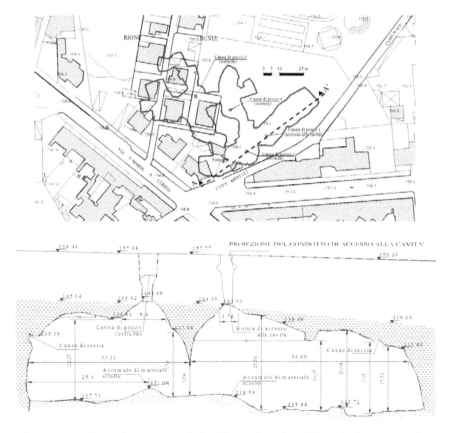

Figure 16. The cavity at Cupa Spinelli. a) plan view. b) vertical cross section.

86

The cavity has been discovered recently because of the opening of a shaft, formerly obstructed by rubble; there are three other shafts, still obstructed. At the site, the thickness of the pyroclastic cover over the tuff is around 15 m. At the surface, a number of apartment houses have been built some decades ago, without even suspecting the occurrence of the cavity.

The tuff cover above the roof of the cavity is relatively thin, and accordingly the stress level is "very high", following the criteria of Table 1. As a matter of fact, a number of blocks fallen from the roof have been found inside the cavity. The cavity has been totally filled with a kind of very fluid lean concrete, poured from the surface through a number of holes. The process has been monitored by a TV camera (fig. 18); the implementation required a careful organisation to transport the enormous amount of material through narrow urban streets. Similar intervention are not uncommon; another case is that of Via Nicolardi (fig. 19).

Other cavities have been adapted to different purposes. For instance, a modern parking has been recently located in one of the oldest and largest cavities in via Domenico Morelli, a short walk to the district's shopping streets. The cave reaches 40 m in height and recalls some of the most defining periods in Naples history: the Greco-Roman cult of Mitra, the 17th century Carmignano Aqueduct, an escape route built for a Bourbon king, a WWII air raid shelter, and a mid XX century waste storage.

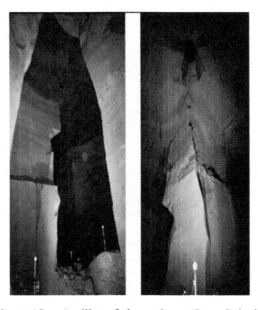

Figure 17.   A pillar of the cavity at Cupa Spinelli.

Figure 18.   Filling the cave at Cupa Spinelli.

Figure 19.   The cavity of Via Nicolardi.

87

The Reichlin quarry (fig. 20), of the rooms and pillars type, is one of the largest; it has been adapted to host a major pumping plant of the city aqueduct. The roof has been consolidated by inclined anchors (fig. 21) or with a r.c. lining founded on piles bored through the debris covering the bottom of the cave. A big cinema was located in the 1950's in another large cavity (fig. 22). Other cavities are worth preserving for public fruition, as the Greek hypogea (Esposito, 2007) and those transformed in worship places (Scotto di Santolo et al., 2013).

A mechanism responsible for a number of accidents in the urban area is that of sinkholes due to erosion. The mechanism is made possible by the existence of underground cavities, overlain by cohe-

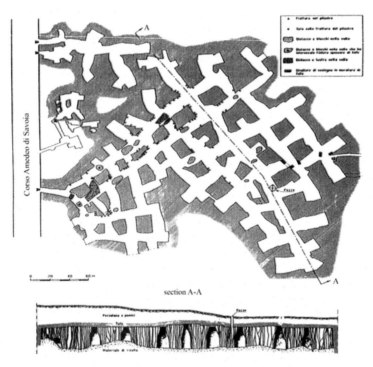

Figure 20.   The Reichlin quarry.

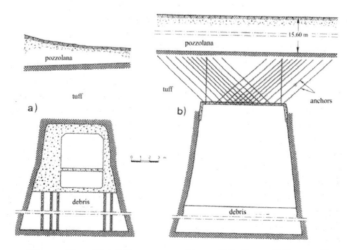

Figure 21.   Reichlin quarry: Consolidation of the roof.

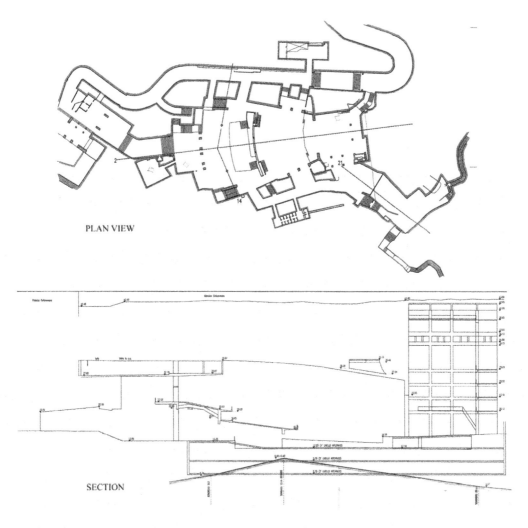

PLAN VIEW

SECTION

Figure 22.   Cinema Metropolitan.

sionless pyroclastic soils, easily erodible. Concentration of rainwater or leaks from aqueducts and sewers may found their way to the underground cavities, with a retrogressive erosion of the soil which is transported into the cavity and creation of a void that progressively enlarges toward the soil surface. The roof of this void becomes progressively thinner, until it reduces, sometimes, to a road pavement. In these conditions a sinkhole eventually occurs, without forewarning, and with a significant potential for damages and casualties. The phenomenon is often initiated by the existence of a vertical shaft, possibly unknown and filled of rubble.

The sinkhole occurred in December 1996 in Via Miano, between the boundary wall of the Royal Garden of Capodimonte and the S. Rocco gorge, may be reported as a typical example. At the surface the hole had an approximately circular shape, with a diameter of around 7 m (fig. 23). The depth was initially around 11 m and increased suddenly to 17 m and then to over 21 m some hours later (fig. 24). The hole had a roughly cylindrical shape, with almost vertical walls, and it developed around a vertical shaft connecting the surface to a deep sewage duct.

Soon after the event the firemen began to excavate at the bottom of the hole, in an attempt to rescue two persons swallowed by the hole.

They excavated under an improvised protection (fig. 25) down to a depth of over 34 m, where they found an underground cavity formed around a collapsed deep sewage duct. The bodies of the two victims were found in that cavity.

Figure 23.   Cross section of the sinkhole in Via Miano.

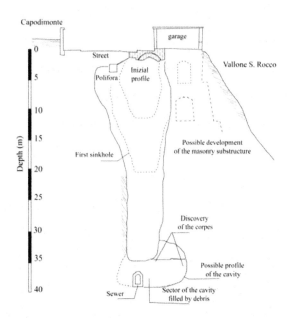

Figure 24.   Cross section of the sinkhole in Via Miano.

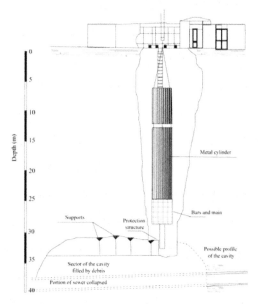

Figure 25.   Provisional protection to retrieve the bodies of the victims.

7   RETAINING STRUCTURES

7.1   *Stability analyses*

A systematic inventory of the retaining walls existing in the urban area has been carried out by a number of small teams, each composed by an engineer and two technicians and operating in a definite zone of the town. For each wall a visual inspection was carried out and a form filled, with the essential information concerning location, potential damages in the case of collapse, eight and thickness at the top, type of structure, drainage provisions, state of conservation of the masonry. 3,300 sections have been documented, belonging to 1,722 walls with a total length of about 110 km.

For 1,300 sections exploratory holes have been drilled to determine the thickness of the wall and the depth of the foundation.

Some of the geometrical data collected are reported in fig. 26 (Evangelista *et al.*, 2002). Almost 50% of the walls have an height H_{ext} lower that 5 m and a ratio between the thickness and the height s/H_{ext} in the range 0.15 to 0.30. The depth of the foundation D ranges between 0.75 m and 1.6 m, with an average value of around 1 m irrespective of the height of the wall.

The stability of a gravity retaining wall depends on a number of factors; besides the geometry of the wall and the properties (unit weight γ, cohesion c', friction angle ϕ') of the backfill and the foundation soil, other significant factors are the inclination of the thrust δ and the passive thrust S_p on the embedded portion of the wall. In order to explore the influence of the various parameters, on the basis of the available evidence some of them have been given a fixed value ($\gamma_{wall} = \gamma_{soil} = 15$ kN/m³; $\phi = 35°$; $D = 1$ m; $S_p = 0$). The cohesion c' of the retained soil and the inclination δ of the thrust have been treated as random variables with mean values M_c, M_δ, standard deviation S_c, S_δ and coefficient of variation $CV_c = S_c/M_c$; $CV_\delta = S_\delta/M_\delta$.

Assuming: $M_c = 10$ kPa; $M_\delta = \phi/2 = 17,5°$; $CV_c = CV_\delta = 0.2$; the mean value M_{FS}, standard deviation S_{FS} and coefficient of variation CV_{FS} of the safety factor of the wall represented in the insert of fig. 27 may be computed (the considered safety factor is the smallest one among those against overturning, sliding and bearing capacity failure).

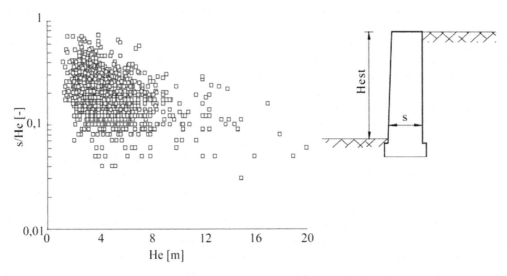

Figure 26.   Geometrical characteristics of the existing retaining walls.

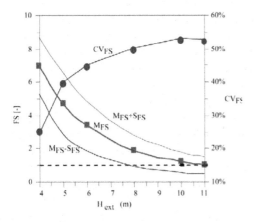

Figure 27.   Influence of different factors on the stability of a retaining wall.

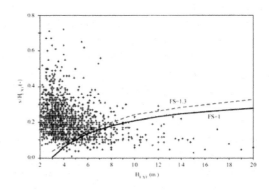

Figure 28.   Stability conditions of the existing retaining walls.

91

The value of CV_{FS} increases with increasing the height of the wall; for $H_{ext} = 4$ m, $CV_{FS} = 24\%$, that is slightly larger than the assumed coefficients of variation of c' and δ. For $H_{ext} = 10$ m and over, CV_{FS} exceeds 50%. In other words the higher is the wall, the larger are the uncertainties in the safety of the wall resulting from the uncertainties in the cohesion and the inclination of the thrust. It may be seen that with the mean value of the parameters a wall with a height $H_{ext} = 11$ m is on the verge of failure ($FS = 1$). Due to the scatter of the parameters, the same wall may be stable with a safety factor equal to 1.5, or the failure may occur for the significantly lower height of 8 m. The scrutiny of a large number of laboratory tests on the pyroclastic soils of Napoli, on the other hand, reveals that the scatter of the cohesion c' is not less than 20% and probably even more.

A second stage of the investigation included a detailed analysis of the stability of 14 selected retaining walls. To this aim a detailed investigation has been carried out for each of them, including the definition of the geometry, the determination of the mechanical properties of the retained and foundation soil, some measurements of the state of stress in the masonry by flat jacks. The stability analyses have been carried out assuming: $\delta = \phi/2$; $S_p = 25\%$ of its maximum theoretical value; cohesion $c' = 0$. With such assumptions only 3 out of the 14 walls should exist, in the sense that for the remaining 11 walls the computed factor of safety is less than unity. Since all the walls are actually stable, the hypothesis of cohesionless soil has been removed. The computed safety factor increases to 1 with a cohesion ranging from 1 to 9 kPa. These findings highlight the very significant role of even a small cohesion in the stability of retaining walls.

Finally, in order to get a first guidance on the stability of the whole population of walls, the following average parameters have been assumed:

- unit weight of the soil $\gamma_{soil} = 15$ kN/m^3;
- unit weight of the wall $\gamma_{wall} = 14$ kN/m^3;
- cohesion of the retained soil $c' = 10$ kPa;
- friction angle of the retained soil $\phi' = 35°$;
- inclination of the thrust $\delta = \phi/2 = 17.5°$;
- depth of the foundation $D = 1$ m;
- passive thrust on the embedded portion of the wall equal to 25% of the theoretical maximum.

The results obtained are reported in fig. 28. The full line represent the geometry of the wall that correspond to the failure conditions ($FS = 1$); the dashed line to a satisfactory safety margin. In the same figure the geometrical properties of the existing walls are reported. With the average conditions assumed, around 20% of the existing walls should have collapsed.

Of course, the existence of these "impossible" walls can be explained by the difference between the average and the local conditions. The area below the $FS = 1$ line, however, represents an attention zone; the walls whose representative points are located there are potentially dangerous and should be checked as soon as possible. On the other hand, it cannot be excluded that even in the zone above the line $FS = 1$ there is some dangerous situation.

The main use of diagrams as the one represented in fig. 28 is that of establishing an order of priority in the interventions.

7.2 *The walls of the vineyard of S. Martino*

The *Tavola Strozzi* (fig. 29) is an oil-on-wood painting of medieval Naples; it is the oldest depiction of the medieval city, dating back to 1472 and very interesting for its amazing details. In the middle of the picture, in a dominant position, there is the S. Martino hill with the Angevin fortress called Belforte and the adjacent monastery. Below the monastery a huge retaining wall is already clearly distinguishable; on the left, a wooded steep slope.

The Carthusian monastery of S. Martino was founded by the Angevins in 1325; the construction of the Belforte started soon after in 1329. The two monuments were probably conceived as a symbol of the religious and military power over the city, and they still impend over the modern Naples.

Figure 29.   The Tavola Strozzi, view of the city of Napoli from the sea, 1470. Museo di S. Martino, Naples.

Figure 30.   Bird's eye view of the S. Martino Vineyard.

Through the centuries the slope of the hill has been gradually converted from an impervious wooded steep crag into a succession of cultivated terraces retained by an impressive system of walls (fig. 30) and provided with a capillary network of drains, channels and water reservoirs. It is still known as "Vigna S. Martino" (S. Martino vineyard).

With the unification of Italy the Monastery became property of the State and seat of the Civic Museum of the city. The underlying vineyard, on the contrary, is at present a private property, although bound to agricultural use and subjected to the public regulations to preserve landscape. In fact, comparing historical maps to the present situation shows that the regulations did succeed in preserving this unique remain in the earth of the city. Since the early XIX century, however, while the vineyard continues to be cultivated, the maintenance of the system of retaining walls and hydraulic structures has been abandoned, and consequently it is steadily deteriorating, with collapse of some walls, landslides, concentrated erosion and with a significant risk for the underlying densely built environment. Fig. 31 reports some examples of the present state of the retaining walls; it is to point out that in the 30 ha of the vineyard there are over 7 km of walls.

The Italian Government has recently made available some resources for the restoration and conservation of the S. Martino hill; a first intervention on the retaining walls is being initiated.

It implies not only the repair of the various forms of degradation affecting them, but also their strengthening to comply with the more demanding requisites of the present seismic regulations. In general, the reconstruction of the outer facing of the walls to repair the effects of the erosion will be carried out with the same tuff masonry of the ancient walls, and cracks will be repaired by local demolition and suture. Some of the counterforts will be underpinned by micropiles, and some new counterforts will be

Figure 31.   Present status of degradation of the retaining walls in the S. Martino Vineyard.

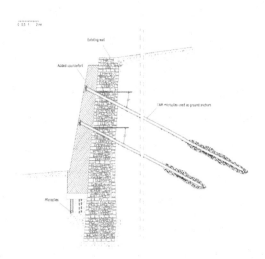

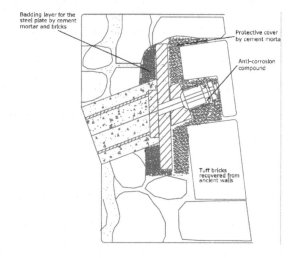

Figure 32.   Reinforcement of a retaining wall with a counterfort anchored and founded on micropiles.

Figure 33.   Connection between the ground anchor and the retaining wall.

added (fig. 32). Finally, in the most critical spots ground anchors will be used, with special care to the connection between the anchor and the masonry (fig. 33).

8 CONCLUDING REMARKS

The mitigation of geotechnical risk in the city of Napoli is a long duration war to drive out the devils and reinstall the angels in the paradise. The spare forces on the battlefield are the Municipality, the University, the (few) good will citizens. They won some battles, and have been ruinously defeated in many other; the final outcome of the war is still completely open. A necessary (though not sufficient) condition for a positive peace will be the availability of adequate resources. Considering the amount needed, the resources shall necessarily come from the central Government; from this point of view, at the time being the forecast cannot be but pessimistic!

In the incipit of his lovely novel "Cannery Row", however, John Steinbeck (1945) writes: *Cannery Row in Monterey in California is a poem, a stink, a grating noise, a quality of light, a tone, a habit, a nostalgia, a dream … . Its inhabitants are, as the man once said, "whores, pimps, gamblers, and sons of bitches," by which he meant Everybody. Had the man looked through another peep-hole he might have said: "saints and angels and martyrs and holy men," and he would have meant the same thing.*

According to Steinbeck, hence, the difference between devils and angels is essentially a matter of peep-hole through whom to look. This gives a chance that the re-appropriation of the Napoli paradise by the angels is not only a nostalgia, a dream.

REFERENCES

Croce B. 2006. *Un paradiso abitato da diavoli.* (G. Galasso, ed.). *Adelphi,* p. 315.
Comune di Napoli 1967. Il sottosuolo di Napoli, relazione della Commissione di studio, p. 446.
Comune di Napoli 1972. Il sottosuolo di Napoli, relazione della seconda Commissione di studio, p. 298 + 5 maps.
Esposito, C. 2007. *Gli Ipogei Greci della Sanità.* Napoli: Oxiana.
Evengelista, A., La Pegna, U. & Pellegrino, A. 1980. Problemi geotecnici nella città di Napoli per la presenza di cavità nella formazione del tufo. *Atti del XIV Convegno Nazionale di Geotecnica, Firenze.* Bologna: Pàtron.
Evangelista, A. & Pellegrino, A. 1990. Caratteristiche geotecniche di alcune rocce tenere italiane. *Terzo ciclo di conferenze di Meccanica ed ingegneria delle Rocce (MIR), Torino,* Vol. 2, pp. 1–32.
Evangelista, A., Cestrone, V., Conte, G., Napoli, M., Lombardi, G., Mandolini, A. & Russo, G. (2002). I muri di sostegno della città di Napoli: inventario, condizioni di sicurezza, interventi di stabilizzazione. *Atti del XXI Convegno Nazionale di Geotecnica, L'Aquila, 11–14 settembre 2002, pp. 579–588. Bologna: Pàtron Ed. ISBN 88-555-26634.*
Melisurgo G. 1889. Napoli sotterranea. Topografia della rete di canali d'acqua profonda. *R. Tipografia Giannini & Figli,* p. 61.
Rasulo, G. 2000. *Il sistema fognario della città di Napoli alle soglie del 2000.* CUEN, Napoli.
Scotto di Santolo, A., Evangelista, L. & Evangelista, A. 2013. The Fontanelle Cemetery: between legend and reality. *Proc. 2nd Inter. Symposium on Geotechnical Engineering for the preservation of Monuments and Historic Sites, TC301,* Naples 2013 (in Press).
Steinbeck J. 1945. *Cannery row* Penguin Group USA (Paperback) edition (1993) ISBN-13: 9780140177381.

Geotechnics and Heritage – Bilotta, Flora, Lirer & Viggiani (eds)
© *2013 Taylor & Francis Group, London, ISBN 978-1-138-00054-4*

The crossing of the historical centre of Rome by the new underground Line C: A study of soil structure-interaction for historical buildings

A. Burghignoli, L. Callisto & S. Rampello
Università degli Studi di Roma La Sapienza, Roma, Italy

F.M. Soccodato
Università degli Studi di Cagliari, Cagliari, Italy

G.M.B. Viggiani
Università degli Studi di Roma Tor Vergata, Roma, Italy

ABSTRACT: This work deals with the soil structure-interaction problems posed by the construction of the third line of Roma underground (Line C), which, in its central stretch, crosses the historical centre of the city with significant interferences with the archaeological and monumental heritage. The paper describes the methodological approach developed to evaluate the effects of tunnelling on the existing monuments and historical buildings, starting from a careful geotechnical and structural characterisation and including the development of reliable geotechnical and structural models. Experts in several disciplines were committed to this multidisciplinary work, ranging from geologists to geotechnical and structural engineers, archaeologists, and professionals working in the field of conservation and restoration of works of art and monuments. The study of the interaction between the construction activities and the built environment was carried out following procedures of increasing level of complexity, from green field analyses, in which the stiffness of the existing buildings was neglected, to full soil-structure interaction analyses, performed in both two—and three-dimensional conditions, accounting for the stiffness of existing buildings and considering possible long-term effects. The paper illustrates the main aspects of this procedure, using the example case studies of the Basilica di Massenzio and of the building of the Amministrazione Doria Pamphili.

1 INTRODUCTION

The growing demand for public and sustainable transport in heavily urbanised areas requires the construction of an increasing number of underground infrastructures. In Italian cities, the use of collective transport is still not fully developed: for example, in Roma public transport covers less than 30% of motorised mobility, compared to 67.7% in Barcelona, 63.3% in Paris, and 47.7% in London. The many constraints and technical challenges associated to the construction of underground infrastructures often lead to high costs and long completion times. This is particularly true in Italy, where many towns are characterised by a high density of population, significant archaeological heritage, and the presence of masonry structures of historical and monumental value, which are particularly sensitive to subsidence induced by excavation. It is therefore often necessary to adopt complex control systems of the excavation process in order to limit the deformations, to devise intense monitoring schemes, and, where necessary, to implement techniques for the protection of the structures affected by the excavation, and these activities result in larger construction costs.

At present, the existing underground network in Rome consists of only two lines, Line A and Line B, intersecting at Termini Central Railway Station. Line B was built during the 1930s by cut-and-cover techniques whereas the bored tunnels of Line A date back only a few decades, as the line was constructed in the 1970s. A northward extension of Line B, or Line B1, has just been completed. The third line of Roma underground, or Line C, will run northwest to southeast across the city centre, for a total length of more than 25 km and 30 stations. The Municipality of Roma approved its preliminary design in October 2002; the tender for detailed design and construction of the line was finally awarded in February 2006, with a value of about 3 billion Euros.

The construction of the south-eastern part of the line is currently under way, while the central stretch of the line, running through the historical centre, from S. Giovanni to Castel S. Angelo, is at a stage of advanced design. This part of the line is particularly challenging because of the geotechnical characteristics of the soil, and the need to minimise the effects on the historical and monumental heritage.

The Authors of this paper had the opportunity to be involved in the project from the very early stages of design, when the Municipality of Roma required the contribution of the University to examine the complex problem of the interaction between the historical and monumental built environment and the construction of the line along Contract T3, from S. Giovanni to Piazza Venezia. A large research team, involving geotechnical and structural engineers from several universities, was set up to respond to the concern of the Superintendence for the protection of the cultural heritage that the construction of this infrastructure could affect adversely the existing monuments. The interaction studies, developed by the team over two years, laid the foundations of a methodological approach that eventually percolated into the contractual requirements included in the tender for the design and construction of Contract T3, between S. Giovanni and Piazza Venezia and Contract T2, between Piazza Venezia and Clodio/Mazzini.

As a matter of fact, the tender included an obligation for the General Contractor to establish an independent technical structure, consisting of a Scientific and Technical Steering Committee (STSC) and several Working Groups (WG), that would undertake detailed studies of the interaction between the line and the monuments that required specific attention. The same technical structure was also to identify the most appropriate and possibly reversible measures for the protection of those monuments that, according to the results of the studies, required intervention.

This was a work of multidisciplinary nature of such complexity as to require the commitment of many scholars, experts in disciplines ranging from geology to geotechnical and structural engineering, archaeology, and conservation and restoration of works of art and monuments.

2 THE CITY ALONG THE TRACK OF LINE C

2.1 *The visible city*

This section explores the visible city along the route of Contracts T3 and T2 of Line C of Roma underground, starting from S. Giovanni Station, where the line intersects the existing Line A and Contract T3 starts (Fig. 1).

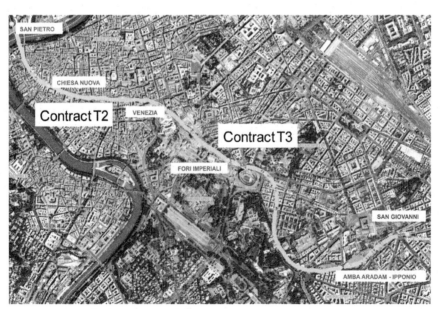

Figure 1.   Line C of Roma underground: Contracts T2 and T3.

98

Emerging at surface from the station and following the Aurelian Walls along Via Sannio, the route encounters Porta Asinaria (Figure 2a), one of the Roman gates in the Aurelian Walls. Following the historic market of via Sannio, the route reaches the first station of Contract T3, at Amba Aradam-Ipponio; under-passing the Aurelian Walls at Porta Metronia, it gets to the Churches of Santo Stefano Rotondo and Santa Maria in Domnica (Figures 2c and d).

Now in the Rione Celio, the line makes a wide curve to the right to get to the Military Hospital and then another wide curve to the left that takes it to the Coliseum, or Anfiteatro Flavio, (Figure 2e), skirted along its north side, to make its way towards Colle Oppio and reaching the second station of the contract, at Fori Imperiali. The route continues along Via dei Fori Imperiali towards Piazza Venezia, in the heart of Roma, passing close to a number of outstanding buildings, including the Basilica di Massenzio (Figure 2f) the Colonnacce of the Foro di Nerva (Figure 2g), the Foro di Augusto, and the Mercato di Traiano.

(a) (b) (c) (d) (e) (f)

Figure 2. The monuments along Contract T3.

From Venezia Station, the route of now Contract T2 continues in the historic centre of the city along the busy Via del Plebiscito, where many important buildings are potentially affected by the line, such as the Church of S. Marco and Palazzo Venezia (Fig. 3a), Palazzo Grazioli, Palazzo Altieri, and the Church of S. Andrea della Valle (Fig. 3b). The route arrives at Largo di Torre Argentina, where there are some of the most ancient monuments of republican Roma, and then, after skirting the Church of Gesù, continues along Corso Vittorio Emanuele, close to a number of historical buildings and churches, including Palazzo Vidoni Caffarelli, Palazzo Braschi, Palazzo Massimo alle Colonne (Fig. 3e), Palazzo della Cancelleria (Fig. 3c), Palazzo Sforza Cesarini, and the Churches of Sant'Andrea della Valle and of San Lorenzo in Damaso. The second station of Contract T2 is at Chiesa Nuova, just in front of the Chiesa Nuova and the Convento dei Filippini (Fig. 3d). Before leaving the baroque bend of the river, the route passes very close to the Antica Zecca (Ancient Mint) (Fig. 3f) and then continues across the Tiber river between Via della Conciliazione and Castel Sant'Angelo.

(a)   (b)

(c)   (d)

(e)   (f)

Figure 3.   The monuments along Contract T2.

2.2 *The underground city and the geological setting*

In its underground crossing of the historic centre, the route of Line C encounters all the main geological features of the site where the city of Roma developed over the centuries; to a certain extent it is these geological features that justify the age and diversity of the surface built environment.

The establishment of the city of Rome was favoured by several auspicious environmental factors, such as the proximity of a navigable river, the presence of abundant spring water, and the nearby existence of quarries for construction materials. For these reasons, the expansion of the city was influenced by the geological factors that have determined its development and that it is useful to recall briefly here.

In the Pliocene and Pleistocene, the sea submerged the area of Roma (see Fig. 4a). The so-called Unit of Monte Vaticano, which represents the basic formation of the subsoil of Rome, was formed in this period; it consists of a deposit of stiff overconsolidated grey-blue clay, nearly 800 m thick.

In the Middle Pleistocene, the uplift of the region and the consequent regression of the sea involved several relocations of the course of the Tiber River, which moved first to the southwest and then to the south. This period is associated to the deposition of a sequence of continental fluvial-marshy deposits consisting of gravel, sand and clay (Fig. 4b).

The volcanic districts of the Colli Albani and Colli Sabatini formed in the Middle-Upper Pleistocene; the associated eruptions and pyroclastic flows changed the course of the Tiber River, moving it to the north, in its present position (Fig. 4c).

Due to the regression of the sea that accompanied the last ice age, the Tiber River cut steadily and deeply into the volcanic and pre-volcanic formations, reaching the Pliocene clays (Fig. 4d); the surviving slabs of volcanic debris anticipate the topography of the present hills of Roma.

During the Holocene, the raising sea level caused progressive filling of the Tiber Valley, which separates the area of Monte Mario—Gianicolo to the west, from the relict volcanic slabs, or the "seven hills", to the east (Fig. 4e).

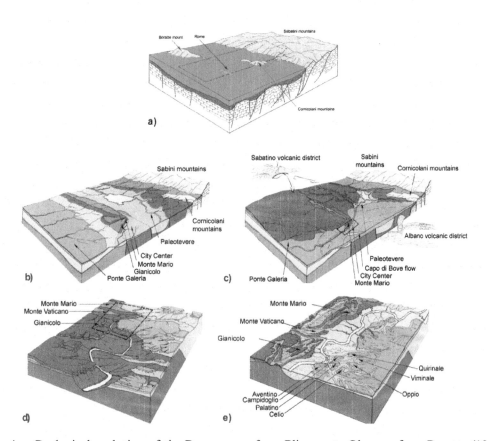

Figure 4.   Geological evolution of the Roman area from Pliocene to Olocene, from Parotto (1990).

In the stretch under examination, Line C runs from S. Giovanni towards Piazza Venezia at an elevation of about 9.5 m above sea level (a.s.l.) or a depth of about 25 m below ground level (b.g.l.) (Fig. 5). The elevation of the track gradually reduces to about 5 m a.s.l. near Piazza Celimontana, corresponding to a depth of 45 m b.g.l, and then increases again to about 0.25 m a.s.l. at Fori Imperiali Station; from this point the line deepens towards Venezia Station, at an absolute elevation of −10 m a.s.l.

In the first part of Contract T3, up to Largo dell'Amba-Aradam, the tunnels run mainly into fine-grained Pleistocene and Holocene soils. After a short passage through the overlying Pleistocene sandy-gravel, the tunnels enter the base Pliocene clay, to emerge again into the overlying formations of sands and gravels, and sands, silts and clays of the Pleistocene, near the Coliseum. At Piazza Venezia there is an abrupt change of the geological environment, as the tunnels enter the Tiber Valley with its Holocene fine and medium grained soils; from here until the end of Contract T3 the tunnels are contained in these type of soils (Fig. 6).

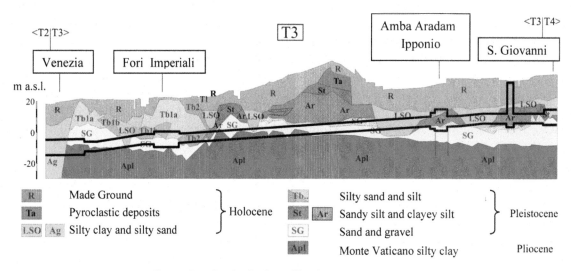

Figure 5.   Geological profile along Contract T3.

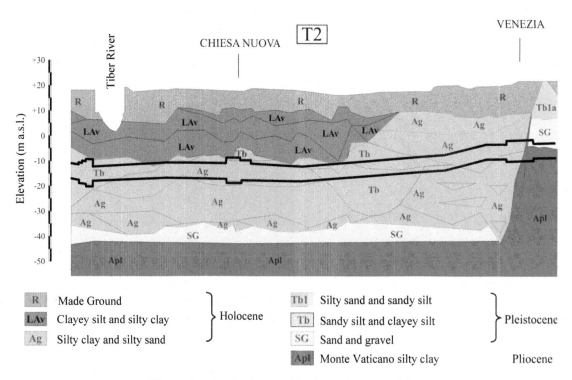

Figure 6.   Geological profile along Contract T2.

102

The pore water pressure distribution along the central stretch of the line is almost hydrostatic, with local small downward gradients; the ground water table is between 10 and 15 m b.g.l. along Contract T3 and between 6 and 8 m b.g.l. along Contract T2.

Between the visible city at surface and the underlying geological environment there are other buried cities, often destroyed and flattened, testifying to the long history of Roma. Figure 7 shows the maps of Rome in 40 BC and 350 AD, while Figure 8 shows the expansion of the city in 1840, before the breakthrough of the urban fabric in the Baroque bend of the Tiber river and in the area of the Roman Forum.

The long history of demolitions and reconstructions taking place over the centuries modified gradually the aspect of the city and, at same time, changed significantly its altimetry, as demonstrated by the relevant thickness of made ground, which, in the area under examination, ranges between 7 and 11 m. The layer of made ground contains the remnants of the ancient city and has an immense archaeological value; any excavation carried out in the historical centre of Roma, for whatever purposes, is bound to raise the interest of the Superintendence for Archaeological Heritage. This requires that preliminary investigations are carried out and authorizations obtained before undertaking the works, often rendering any estimate of their duration and overall costs extremely uncertain.

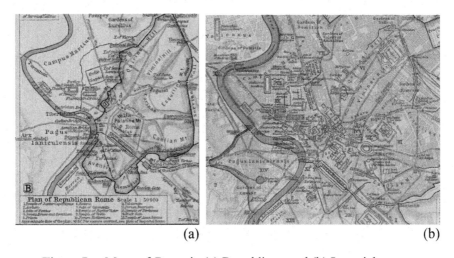

(a)   (b)

Figure 7.   Maps of Rome in (a) Republican and (b) Imperial ages.

Figure 8.   Map of Rome in 1841.

2.3 Tunnelling method

Construction of Line C includes excavation of two running tunnels with a diameter of 6.7 m, using Earth Pressure Balance (EPB) Tunnel Boring Machines (TBM).

The essence of an EPB machine for tunnelling in soft ground is provision of substantial support to the excavated face at all times, thereby controlling ground movements. Face support is provided by the cutterhead, powered by a drive motor, contained within the circular steel skin of the TBM. The soil, excavated by the rotating cutter wheel, passes into the excavation chamber immediately behind the cutterhead. The excavated material is removed from the excavation chamber by an auger conveyor (cochlea).

The screw conveyor plays an important role in the excavation process. As the machine advances through the ground, the excavated soil enters the pressurised head chamber. The soil is extracted from the head chamber and flows along the screw conveyor to the discharge outlet, where it is discharged at atmospheric pressure onto a conveyor belt. The rotational speed of the screw and the opening of the upper auger conveyor driver influence the soil flow rate and pressure gradient along the screw conveyor. Controlling the rate of soil discharge and the pressure gradient along the screw conveyor regulates the head chamber pressure supporting the tunnel face. If the machine advances steadily, a reduction in the screw conveyor extraction rate will cause an increase in pressure in the excavation chamber; correspondingly, an increase in extraction rate will result in a reduction in chamber pressure.

The factors influencing the chamber pressure during excavation are complex but the details of the screw conveyor operation are of particular relevance. It is important that the extraction of the soil is well controlled, synchronised with the speed of excavation, and that the soil mixture is converted to a low shear strength paste by suitable soil conditioning. This is achieved by injecting conditioning agents, most commonly foams or polymers, into the cutterhead to mix with the soil during the excavation process. The parameters that have to be selected for the soil conditioning comprise the type of product (water, bentonite, polymer, foam or any combination of these) as well as their quantities.

The tunnel is lined with reinforced concrete segments, which are positioned under atmospheric pressure by means of an erector arm in the rear area of the shield. As the tail skin leaves the tunnel lining, grout is injected under pressure to fill the annular void between the extrados of the segmental lining and the excavated ground. Tail skin seals prevent the grout from entering the TBM. Appropriate control of the excavation parameters and timely and effective tail skin grouting with mixtures of low permeability are the determining factors to obtain very low values of volume loss. With this type of TBMs the volume loss is generally lower than 1%.

3 EFFECTS OF TUNNELLING AT SURFACE

3.1 Subsidence

Urban tunnels are typically shallow and, as such, interact with the built environment.

In green field conditions, *i.e.*, in the absence of structures, the settlement trough induced by tunnelling has a characteristic shape (Fig. 9.). The available field evidence indicates that the surface settlement trough may be described by a Gaussian distribution curve in a section transversal to the tunnel axis, at sufficient distance from the face to assume plane strain conditions, and by a cumulative probability function in the longitudinal direction (Peck, 1969, O'Reilly and New, 1982).

The transverse surface settlement trough can be written in the form:

$$w = w_{max} \cdot \exp\left(\frac{-y^2}{2 \cdot i^2}\right) \tag{1}$$

where w is the vertical displacement a distance y from the tunnel axis, w_{max} is the maximum vertical displacement, controlling the magnitude of the subsidence, and i is the distance of the point of inflection of the Gaussian curve from the tunnel axis, defining the width of the settlement trough ($\cong 6i$).

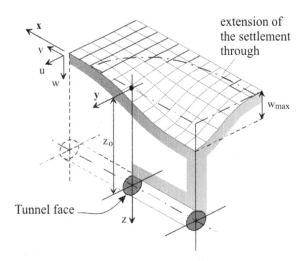

Figure 9. Subsidence trough in greenfield conditions.

The volume of the surface settlement trough per unit length of advancement, V_S, can be obtained by integration of the Gaussian curve:

$$V_S = \sqrt{2\pi} \cdot i \cdot w_{max} \qquad (2)$$

In saturated fine-grained soils and undrained conditions, the volume of the settlement trough at surface is equal to the over-excavated volume at the tunnel. In coarse-grained soils this is not the case, as the volume of the settlement trough at surface may be less than the over excavated volume, due to dilatancy of dense sand on shearing. Whatever the soil type, it is convenient to define the volume loss, V_L, as the (percentage) ratio of the volume of the surface settlement trough and the excavated nominal volume of tunnel, i.e., for a circular tunnel of diameter D:

$$V_L = \frac{4V_S}{\pi D^2} \qquad (3)$$

In the present study, values of the volume loss between 0.5% (contractual requirement) and 1.0% (worst case scenario) were used. Combining equations (2) and (3), it is possible to express w_{max} as:

$$w_{max} = \frac{V_L \cdot \pi D^2 / 4}{\sqrt{2\pi} \cdot i} \qquad (4)$$

For relatively deep tunnels, i.e., tunnels with a cover at least equal to one diameter, the value of i at surface is proportional to the depth of the tunnel axis, z_0 (O'Reilly e New, 1982):

$$i_0 = K \cdot z_0 \qquad (5)$$

through a width parameter, K, that depends on the type of soil between tunnel crown and surface, and takes values in the range 0.2–0.3 for sands above the water table, 0.4–0.5 for stiff clays, and 0.6–0.7 for soft clays (Ribacchi, 1993). In the present study, values of $K = 0.4$ to 0.5 were assumed at ground surface.

For ground movements at depth it is typically assumed (O'Reilly & New, 1982; Grant & Taylor, 2000) that the shape of the transversal settlement trough is still Gaussian so that eq. (1) can be used to compute the profiles of vertical displacements at any depth z below ground surface, once w_{max} and the variation of $i = K(z) \cdot (z_0 - z)$ with z are defined.

Several relationships to evaluate i at depth have been proposed in the literature, such as those by Mair et al. (1993), for stiff clays:

$$i(z) = 0.5z_0 - 0.325z \qquad (6)$$

105

or by Moh et al. (1993), for coarser grained soils:

$$i(z) = bD \left(\frac{z_0 - z}{D} \right)^m \tag{7}$$

in which b and m are parameters that depend on the characteristics of the soil. The present study adopted the expression by Moh et al. (1996) with values of the exponent $m = 0.6$ that is in between the values of 0.4 and 0.8 recommended for silty-sands and silty-clays, respectively. Parameter b was obtained from the value of i at surface:

$$i(z = 0) = i_0 = b \cdot D \cdot \left(\frac{z_0}{D} \right)^m \tag{8}$$

or:

$$i(z) = i_0 \cdot \left(\frac{z_0 - z}{z_0} \right)^m \tag{9}$$

To evaluate the horizontal components of displacements it was assumed that the vectors of displacement in the plane orthogonal to the tunnel axis point towards the centre of the tunnel (O'Reilly & New, 1982), and, therefore, the horizontal displacements, u, can be computed from the vertical displacements as:

$$u(y,z) = \frac{y}{z_0 - z} \cdot w(y,z) \tag{10}$$

In the longitudinal direction, the vertical displacements may be computed as (O'Reilly & New, 1982, Attewell & Woodman, 1982):

$$w(x,y,z) = w_{\max} \cdot \exp \left(\frac{-y^2}{2 \cdot i(z)^2} \right) \cdot F \left(\frac{x}{i(z)} \right) \tag{11}$$

Where $F(x)$ is the a cumulative probability function:

$$F(x) = \int_{-\infty}^{x} \frac{1}{\sqrt{2\pi}} \cdot \exp \left(\frac{-t^2}{2} \right) \cdot dt \tag{12}$$

The displacement field described so far is relative to the short term, resulting from the progress of excavation and the installation of the lining. For tunnels excavated in fine-grained soils, settlements can increase with time due to the change in the hydraulic conditions at the tunnel internal boundary, where pore pressures are equal to zero. Reviewing the long-term settlements measured in clays after tunnel construction, Mair and Taylor (1997) concluded that the major factors influencing their development are the initial pore water pressure distribution, the excess pore water pressure generated by tunnel construction, the compressibility of the clay and, most significantly, the ratio of the permeability of the tunnel lining to that of the soil. Mair (2008) discussed example case studies, showing that the tunnel lining should be regarded as a porous material.

Wongsaroy (2005) identified the key factors affecting possible long-term effects and introduced two dimensionless variables:

$$DS = \frac{w_{\max} - w_{\max,\text{imp}}}{w_{\max,\text{per}} - w_{\max,\text{imp}}} \tag{13}$$

$$RP = \frac{k_l}{k_s} \frac{z_0 - 0.5D}{t} \tag{14}$$

in which w_{max}, $w_{max,imp}$, and $w_{max,per}$ are the actual maximum long-term settlement and the maximum long term settlements for a totally impermeable and totally permeable lining, k_l and k_s the coefficients of permeability of the lining and the soil, and t the thickness of the lining.

The results of the parametric study by Wongsaroy (2005), shown in Figure 10, indicate that the lining can be considered as perfectly impermeable if the relative permeability of the lining to the soil is less than 0.1 and wholly permeable if the relative permeability is larger than 100, which can be used to orient design of lining and clogging injections.

Despite the fact that the long-term settlements are generally larger than in the short term, their effects on the existing structures are generally limited because they tend to increase the width of the settlement profile thus reducing differential settlements, curvatures and distortions.

3.2 *Interaction with structures at surface*

The displacement field generated by the excavation of the tunnel propagates to the surface, thus interacting with the existing buildings located within the settlement trough.

The buildings that are closer to the tunnel axis tend to assume a deformed configuration with an upward concavity (*sagging*), while the buildings that are far from the tunnel axis, beyond the point of inflection of the settlement profile, tend to assume a deformed configuration with a downward concavity (*hogging*). The presence of structures modifies the shape of the settlement trough, typically reducing distortions and curvatures, to a greater or lesser extent depending on their stiffness. If the values of curvature exceed given thresholds, the structures will experience damage of varying severity, all the way from simple aesthetic to severe structural damage. In general, buildings suffer more from the hogging than from the sagging mode of deformation (see Figure 11). This due to the restraining effect produced by the ground and by the foundations in the lower part of the structure, while in the hogging mode, which is generally missing in the upper parts of the structure.

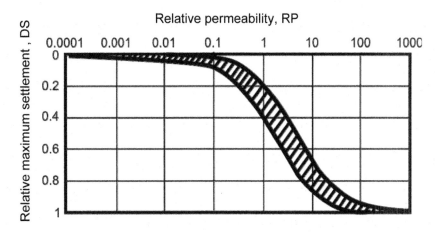

Figure 10. Effects of soil-lining relative permeability on the long-term maximum settlement (Mair, 2008).

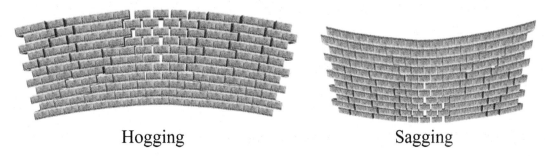

Hogging   Sagging

Figure 11. Hogging and sagging in masonry walls.

Older or ancient buildings are often already affected by damage of various origin, mainly connected to self weight, manifesting with existing crack patterns; new damage can arise as intensification of existing cracks but may also affect parts of the building that had not been previously affected by damage due to self weight or other causes. All these factors must be taken into account in the assessment of expected damage to old structures.

4 EXPECTED DAMAGE

Developing a rational and completely objective evaluation of the risk of damage to historical structures due to tunnelling and deep excavations is not an easy task. The wealth of historical, artistic and monumental buildings potentially affected by the construction of Line C of Roma underground made it necessary to reconsider the concept of damage and its identification and classification. Following the approach by Burland *et al.* (1977), in this study risk is associated with a potential degree of damage, so that the judgment on a low or high level of risk is reduced to the evaluation of a low or high degree of damage. This justifies the establishment of a classification of damage, illustrated in the following section, connecting its severity to representative indicators.

4.1 Classification of damage

Burland *et al.* (1977) proposed a classification of damage to masonry buildings. The classification is based on the ease of repair and provides information on the visible effects of damage, such as crack widths.

The classification in Table 1 identifies six categories of damage of increasing severity, from negligible to very severe, which can be grouped into three broader classes: damage affecting appearance or visual aesthetics (categories 0, 1, and 2), function (categories 3 and 4), and stability (category 5). It is probably worth to recall that different causes, such as thermal effects or seasonal oscillations of the groundwater table, may result in damage up to category 2.

Table 1. Damage classification, after Burland *et al.* (1977).

Category of damage	Normal degree of severity	**Description of typical damage** (ease of repair in bold type) Note: *Crack width is only one factor in assessing category of damage and should not be used on its own as a direct measure of it*
0	Negligible	Hairline cracks less than about 0.1 mm wide.
1	Very slight	**Fine cracks that are easily treated during normal decoration.** Damage generally restricted to internal wall finishes. Close inspection may reveal some cracks in external brickwork or masonry. Typical cracks widths up to 1 mm.
2	Slight	**Cracks easily filled. Redecoration probably required. Recurrent cracks can be masked by suitable linings.** Cracks may be visible externally and **some repointing may be required to ensure weather-tightness.** Doors and windows stick slightly. Typical crack widths up to 5 mm.
3	Moderate	**The cracks requires some opening up and can be patched by a mason. Repointing of external brickwork and possibly a small amount of brickwork to be replaced.** Doors and windows sticking. Service pipes may fracture. Weather tightness often impaired. Typical crack widths are 5–15 mm or several > 3 mm.
4	Severe	**Extensive repair work involving breaking-out and replacing sections of walls, especially over doors and windows.** Windows and door frames distorted, floor sloping noticeably. Walls leaning or bulging noticeably, some loss of bearing in beams. Service pipes disrupted. Typical cracks widths are 15–25 mm but also depends on the number of cracks.
5	Very severe	**This requires a major repair job involving partial or complete rebuilding.** Beams lose bearing, walls lean badly and require shoring. Windows broken with distortion. Danger of instability. Typical crack widths are greater than 25 mm, but depends on the number of cracks.

Evaluating the crack pattern induced by tunnelling or the evolution of the pre-existing crack pattern is a very complex task, because of the difficulty of modelling adequately the structural behaviour of masonry. Assuming that cracking of the walls results mainly from tensile strains, Burland & Wroth (1974) proposed that the value of tensile strain might be considered as the key indicator of structural damage, thus opening the way for much simpler assessment of the structural behaviour.

Following this idea, and considering the indications by Boscarding & Cording (1989) on the values of limiting horizontal tensile strain, Burland (1995) extended the previous classification of damage, to provide so-called "interaction diagrams" in which the identification of the category of damage results from the combination of the computed horizontal tensile strain and deflection ratio, see Figure 12. In this manner, assuming that a masonry wall behaves like a simple beam with bending and shear stiffness (Timoshenko beam), the category of expected damage can be evaluated in a relatively simple manner. This approach has been used extensively in the design phase of the Jubilee Line Extension in London, as an intermediate step of a process of damage assessment that included a preliminary estimate, based on the green field deformations, followed by more advanced analyses in which soil-structure interaction was taken into account.

The extent to which the approach outlined above can be applied to the buildings of the centre of Roma and, more generally, the definition of a strategy for the preservation of their historical, artistic and monumental value are difficult issues that require appropriate consideration.

First of all, the construction of an infrastructure such as Line C of Roma underground, with a significant social impact, can only derive from the result of a careful cost/benefit analysis demonstrating that the benefits prevail. Therefore, society as a whole is called to carry the burden of the problems that may result from its construction and operation, provided, of course, that they are predicted in design and contained within acceptable limits.

The definition of "acceptable limits", however, may constitute a problem. When the potentially affected buildings are of outstanding value, an extreme position may be that of considering unacceptable any new work interacting with them, unless specifically intended for their preservation or valorisation. A striking example of this attitude is provided by the archaeological excavation required to expose the Arco dei Ladroni near Basilica di Massenzio, which required the installation of a loud and aesthetically questionable structural safeguard measure to minimise the risk of collapse of part of the monument.

If, on the one hand, the protection of the monumental heritage requires that the absence of damage due to the works be guaranteed, on the other hand this cannot mean the absence of any interaction between the new works and the old buildings. In the case at hand, more stringent constraints had to be introduced to ensure the best protection of the monumental heritage and a consistent and scientifically

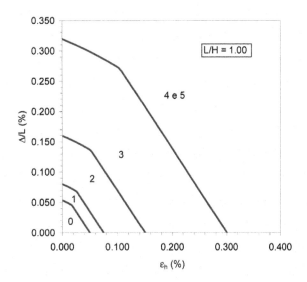

Figure 12. Interaction diagram and iso-damage zones for $L/H = 1.0$, after Burland (1995).

sound methodological approach developed to evaluate how the construction of the line would affect the existing historical buildings.

To this end, the analysis of the interaction between construction activities and built environment was carried out following procedures of increasing complexity, level of detail and accuracy.

4.2 *Methodological approach*

The evaluation of the expected damage was carried out with reference to the classification of damage by Burland (1995) using analyses of different levels.

At a first level, simplified analyses were performed computing surface and near-surface displacements using the semi-empirical methods described above and neglecting the stiffness and the weight of the buildings (Attewell & Woodman, 1982, Attewell *et al.*, 1986).

The resulting displacement field was applied by the structural engineering group to a 3D linear elastic finite element model of the structure under examination. Both the geotechnical and structural engineering groups carried out independent evaluations of potential damage to the buildings. To account for the historical value of the buildings, the values of limiting tensile strains adopted by the geotechnical engineering group to identify the damage categories were lower than those proposed originally by Boscardin & Cording (1989). Based on the outcome of the geotechnical and structural evaluations, the study ended if the damage was deemed negligible, or continued to a higher level of complexity (Level 2).

At this second stage, the interaction between the tunnels and the historical buildings was studied through 2D or 3D finite element analyses that considered soil-structure interaction adopting a simplified description of the mechanical behaviour of the buildings. The computed displacement field, accounting this time for the stiffness and weight of the building, was applied again to the structural model and damage was re-evaluated, independently, by both groups. Depending on the computed results, either damage was deemed acceptable, or prospective remedial techniques were indicated.

5 SOIL STRUCTURE INTERACTION ANALYSIS

Soil structure interaction studies require the definition of a computational strategy yielding the same level of approximation for both geotechnical and structural analyses. This requirement stems from two sets of problems. The first is connected with the specialization of the available numerical codes, which often makes it impossible, or at least very difficult, the task of tackling structural problems with "geotechnical" codes and vice versa. A second problem derives from the skills and sensitivity required to use effectively specialised software for the structural and geotechnical aspects. This suggested to carry out independent geotechnical and structural analyses, but also to define a procedure that would permit to study the interaction between the buildings and the ground with the same level of approximation for the geotechnical and structural aspects.

The soil-structure interaction analyses adopted a simplified mechanical description of the buildings, obtained representing the structure as an equivalent solid with simpler geometry and appropriate physical and mechanical properties. This equivalent solid has the same shape in plan as the real building it is meant to represent, is fully embedded into the soil and is assumed to behave as a linear elastic material. The equivalent solids are introduced in the geotechnical finite element analyses with equivalent stiffness and weight that reproduce those of the complete buildings. In this manner, the geotechnical analyses produce a displacement field that accounts for the stiffness and weight of the building and that can be eventually applied to the complete structural model for a final evaluation of the effects induced by tunnel construction.

Strictly, the assumption of linear elasticity for the equivalent solid is valid only when the structure undergoes small distortions. When the mechanical non-linearity of the real structure plays a major effect, in principle the procedure may require one or more iterations. However, the results of the analyses have proved that generally this is not the case, at least for relatively small volume loss.

5.1 Geotechnical models

To limit the computational effort, particularly for three-dimensional analyses, in a first stage of the study the soil was modelled as linearly elastic-perfectly plastic, with a Mohr Coulomb yield criterion. Subsequently, to overcome the limits of this constitutive law, the mechanical behaviour of all soils was described using an elastic—plastic rate independent constitutive model with isotropic hardening that his capable of reproducing the main features of the mechanical behaviour of soils, namely the *Hardening Soil* model (Schanz *et al.*, 1999), available in the library of the finite element codes Plaxis and Tochnog that were used for the numerical analyses.

In the constitutive model, the elastic behaviour is defined by isotropic elasticity using a stress dependent Young's modulus, while plasticity is governed by a deviatoric and a volumetric yield surface with independent isotropic hardening laws, related to deviatoric and volumetric plastic strains, respectively. The flow rule is associated for states lying on the volumetric surface, while a non-associated flow rule is used for states on the deviatoric surface.

Compared to linear elastic-perfectly plastic models, the use of *Hardening Soil* results in a significant improvement of the analyses because it considers a non-linear (hyperbolic) stress-strain relationship; takes into account the stiffness variation on plastic loading and unloading; considers the dependence of the stiffness on the current stress state and on the over-consolidation ratio; predicts the occurrence of plastic strains from the onset of deviatoric loading, even for overconsolidated states, thanks to the separation of the volumetric and deviatoric yield surfaces.

5.2 Structural models

In principle, the dual objective of modelling accurately the mechanical behaviour of masonry and of identifying potential damage, even at the level of the structural detail, would imply the use of three-dimensional non-linear numerical analyses. As the buildings potentially affected by construction of Contracts T2 and T3 and deemed worthy of attention are numerous, it was not possible to carry out this type of analysis for all of them and this approach was followed only in some specifically selected cases.

A reasonable balance between the computational restrictions and the need to achieve the objectives of the interaction studies, was offered by an application of damage mechanics, that treats masonry as a continuum and introduces damage as a tensorial variable. In a simplified isotropic model proposed by Mazars (1984) for concrete, damage was represented by two independent scalar variables, describing the behaviour in extension and in compression, see Figure 13. This was the reference model adopted at the early stage of the study, within a research contract between the Municipality of Roma and the University (STA-DISG, 2003), considering only damage in extension as the most relevant for masonry walls. In successive studies, the same model was also used to evaluate damage from the results of linear analyses.

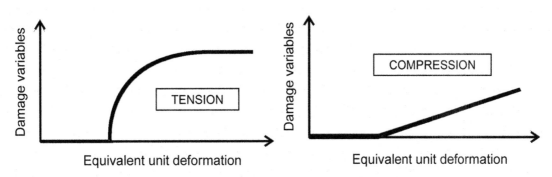

Figure 13.   Damage variables, after Mazars (1984).

The results of a parametric study of the behaviour of masonry walls with rectangular openings and different ratios of the void to masonry areas, shown in Figure 14, demonstrated that linear and non-linear models provide very similar results in sagging, while in hogging the results are in reasonable agreement only up to values of deflection ratio, Δ/B, of 1×10^{-4}.

The results of this study suggested that it was possible to use linear models in all those cases in which Level 1 analyses indicated negligible damage, limiting the use of non-linear models to only a few cases, with the main function of support of the results of the linear analyses.

5.3 Equivalent solids

The technical literature contains several examples of equivalent solids, typically used to model masonry structures in numerical geotechnical analyses. After the work by Burland & Wroth (1974), who used the simple Timoshenko beam to obtain their classification of damage, other Authors (Finno *et al.*, 2005; Pickhaver, 2006) have regarded the façades of masonry buildings as equivalent beams of appropriate shear and bending stiffness.

In the literature, it is generally recognised that the simple assumption of linear elastic isotropic behaviour does not describe the behaviour of real structures adequately, and that it may be necessary to adopt values of the ratio of the Young's and shear modulus, E/G, outside the permitted range of isotropic elasticity, depending on the type of building. For a linear elastic isotropic material with a Poisson's ratio $\nu = 0.3$, E/G is equal to 2.6; however, for reinforced concrete frame structures, with a relatively high shear deformability, it is advisable to adopt values of E/G as high as 12.5, while for masonry structures the recommended values of E/G may be as low as 0.5.

Voss (2003) extended the work of Burland & Wroth (1974) to propose an expression for the deflection ratio, Δ/L, in which the ratio E/G appears explicitly, together with the distance of the neutral axis from the bottom of the beam, normalised by the beam's height, see Figure 15.

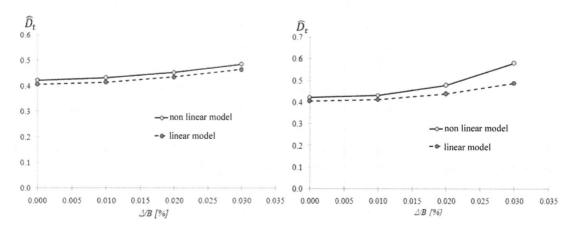

Figure 14.   Parametric study of damage in masonry wall.

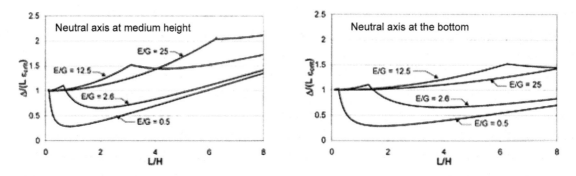

Figure 15.   Deflection ratio as a function of E/G for an elastic beam, adapted from Voss (2003).

112

In the approach by Finno *et al.* (2005), the equivalent beam is identified assuming that the floors contribute to the bending stiffness of the building, while the walls contribute to its shear stiffness. The authors end up with a laminated beam model, see Figure 16, in which the different layers correspond to floors and walls.

In the work by Pickhaver (2006) the façades of the building are modelled as elastic beams, whose behaviour depends on the percentage of openings and on the mode of deformation.

Based on the results of an extensive parametric study of the behaviour of rectangular masonry façades with different percentage of openings, carried out with linear finite element analyses, Pickhaver (2006) demonstrated that: (i) the Timoshenko beam does not provide an adequate representation of the behaviour of the walls and (ii) there is a critical value of the ratio L/H, between the length and the height of the façade, below which the stiffness depends essentially on the ratio L/H and above which the stiffness depends essentially on the percentage of openings, see Figure 17. The Author proposed also a procedure to assess the equivalent values of the area and inertia of the cross section of the beam, to take into account the above-mentioned factors.

In this study, the issue of the identification of the equivalent solid was addressed anew. In particular, it was assumed that the actual and the simplified structure can be considered equivalent if they show the same response to a given perturbation. For the present problem, the perturbation consists of imposing the vertical displacements computed in the green-field analyses at foundation level, and the corresponding response is the distribution of the nodal forces at the same level. The Young's modulus of the equivalent soil is found iteratively to produce a distribution of nodal forces at foundation level that matches the distribution computed using a complete structural model of the building, see Figure 18.

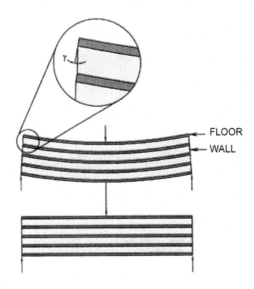

Figure 16.   Scheme of laminated beam (Finno, 2005).

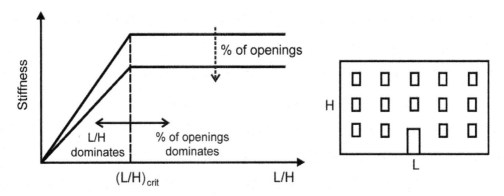

Figure 17.   Effect of L/H and of percentage of openings on the stiffness of masonry walls (after Pichaver, 2006).

113

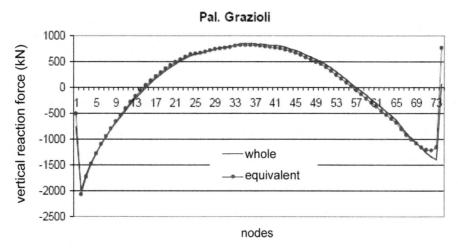

Figure 18. Example of calibration of the equivalent solid based on node reactions.

The thickness of the equivalent solid has been generally chosen so as to occupy, in the mesh used for geotechnical soil-structure interaction analyses, the same space of the foundations of the real building.

6 BACKGROUND NOISE

In the context of this study, the term "background noise" signifies the variation in time, before construction, of all those physical quantities, such as surface movements and pore water pressures, which will be modified by the construction of the line. In the case under examination, it is very important to examine the background noise in terms of building settlements, as the background displacements can be of the same order of magnitude as those induced by construction.

A number of factors, such as the geological environment in which the city developed, the geotechnical characteristics of the foundation soils, the presence of the Tiber and the periodic fluctuations of its water level, and the daily and seasonal thermal cycles, combine to generate movements of the buildings.

For instance, automatic total station monitoring of the Basilica di Massenzio revealed that the building experiences every day displacements of the order of millimetres, due to the daily thermal cycle (Figure 19).

The progressive settlements of Palazzo di Giustizia, before the reinforcement of its foundations, could be attributed to fluctuations in the hydrometric regime of the Tiber River (Calabresi et al., 1980), see Figure 20. In fact, the same cause is responsible for the movements of a large number of buildings of different type and size, located along the river. As shown in Figure 20, the fluctuations in the hydraulic head in the layer of gravel and sand immediately above the base Pliocene clays follows the hydrometric level fluctuations in the Tiber River; this creates deformations of the overlying compressible soils and is responsible for the extension of the area of influence of the Tiber hydrometric level.

Using a monitoring technique based on the creation of maps of surface deformation from SAR (Synthetic Aperture Radar) images, it is possible to obtain a very accurate representation of the elevation of numerous *permanent scatterers, i.e.,* details that always reflect the radar signal in the same way during different passages of the radar, such as rooftops or terraces of buildings. The technique can potentially measure millimetre-scale changes in deformation over spans of days to years.

Figure 21 shows an overall view of the permanent scatterers in the city of Roma; these are represented as small circles ranging in colour from blue to red depending on the measured value of the average annual rate of settlement. Consistently with what just discussed, it is possible to identify a strip along the Tiber characterised by higher than average values of annual settlement rate and some critical areas, typically corresponding to problematic foundation soils, where the buildings show significant ongoing settlements.

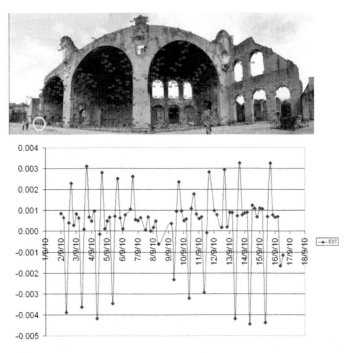

Figure 19.　Effect of the daily thermal cycle on the displacements of Basilica di Massenzio.

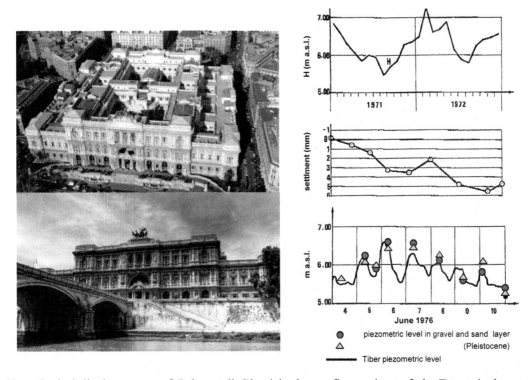

Figure 20.　Vertical displacements of Palazzo di Giustizia due to fluctuations of the Tevere hydrometric level (after Calabresi *et al.*, 1980).

Taking a closer look to the buildings of historical centre, potentially affected by the construction of Contracts T2 and T3, the data in Figure 22 indicate that the foundation soils in the baroque bend of the Tiber are more compressible than the oldest soils between Piazza Venezia and Coliseum. Even at the scale of individual or adjoining buildings, SAR data show noticeable average settlement rates, up to 0.2 mm/year, and seasonal oscillations of the order of the millimetre, see Figure 23.

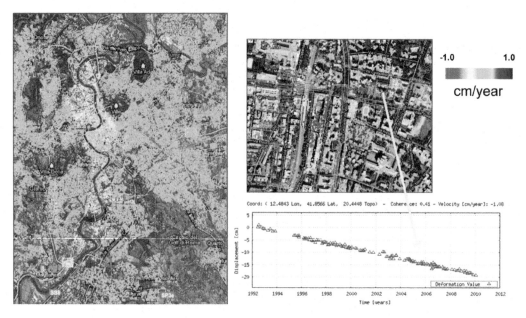

Figure 21.   SAR survey: Distribution of average annual settlement rates for the whole urban area.

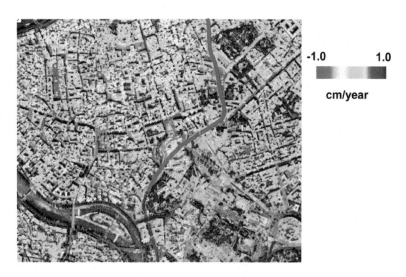

Figure 22.   SAR survey: Distribution of average annual settlement rates for Contracts T2 and T3.

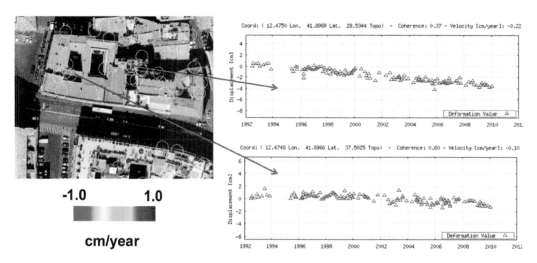

Figure 23.   SAR survey: Distribution of settlement rates at the single-building level.

116

The magnitude of the measured background noise illustrated above suggests that it is probably pointless to try to predict the effects of the construction of the line with accuracies smaller than about one millimetre and to attach particular relevance to displacements of the order of a few millimetres in the assessment of potential damage.

7 EXAMPLES

7.1 *Basilica di Massenzio*

The Basilica di Massenzio is a very interesting example to illustrate the approach that was adopted in the study. In fact, this is a particularly heavy structure, with significant contact stress transferred to the soil by the foundations that consist of simple extensions of the bearing walls, with a very limited widening. Another reason of interest of this case study derives from the variety of solutions that were considered in the design of the tunnels and of the station shaft, that made it necessary to carry out several interaction analyses.

7.1.1 *History*

Construction of the Basilica began on the northern side of the forum under emperor Massenzio in 308, and was completed in 312 by Costantino I, after his defeat of Massenzio at the battle of Ponte Milvio.

In its original configuration, the building consisted of a central nave, covered by three vaults on four large piers and ending in an apse at the western end, and two flanking aisles spanned by three semi-circular barrel vaults perpendicular to the nave. Excluding the apses, the building occupied a rectangular area of about 80×60 m^2 (Fig. 24).

Structural failures of the building occurred as early as during construction, probably during the interruption of the works for the civil war between Costantino I and Massenzio. Subsequent reinforcements included the construction of buttresses and contrast arches.

The perimeter walls of the Basilica, as well as the internal baffles, consist of two facings of clay bricks (*opus testaceum*) and a core of Roman conglomerate of lime and pozzolana (*opus caementicium*) including aggregates of different materials. The vaulted structures as well as the foundations are made in *opus caementicium*. The ceilings of the barrel vaults show advanced weight-saving structural skill with octagonal ceiling coffers.

In the fourth and fifth centuries the Basilica underwent several modifications, including the creation of the apse on its northwestern side, and the construction of a retaining wall to support the Velia Hill. In the sixth century, the Basilica had been already abandoned.

Figure 24.   Map of Basilica di Massenzio after the restoration works of IV and V centuries.

Subsequent repeated stripping and invasive crafts established on the site led to a progressive deterioration of the monument; the south and central sections were probably destroyed by the earthquakes of 847 and 1349. In the following centuries, what was left of the monument was affected by progressive accretion and improper utilisation, including use as stable and riding school (Fig. 25).

The first excavations to restore the Basilica to its original level began in the nineteenth century and in 1932 the excavation works to remove the Velia Hill and make room for the new Via dell'Impero were carried out; exposed by these works, the structure revealed the presence of an extensive pattern of cracks and significant damage in the two vaults and the apse.

In the 1960s, Musumeci reconstructed the destroyed dome of the apse in reinforced concrete; the present aspect of the Basilica is that reported in the photo of Figure 26.

Finally, Figure 27 illustrates the construction stages of the Basilica, which were used for its structural modelling.

7.1.2 *Ground conditions*

The geotechnical characterization of the foundation soils of the Basilica was undertaken using the results of site and laboratory tests carried out during several geotechnical investigation campaigns. Figure 28 shows a plan view of the Basilica and of the running tunnels together with the location of

Figure 25.   View of the monument in 1865.

Figure 26.   View of the monument in 2000.

118

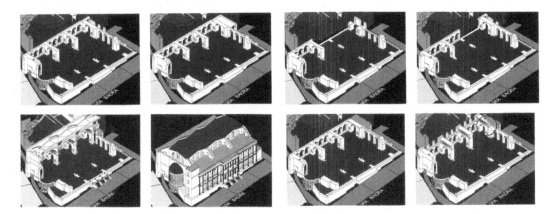

Figure 27.   Construction stages of Basilica di Massenzio.

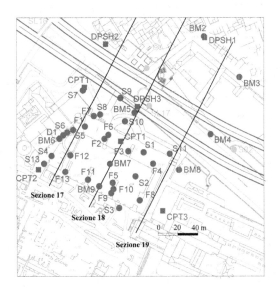

Figure 28.   Geotechnical investigations in the area of Basilica di Massenzio.

all the boreholes and *in situ* tests. The same figure also shows the position of three reference sections used for the interaction analyses, while Figure 29 details the stratigraphy along the central transversal section, intersecting the apse.

In the preliminary design, the tunnels had an external diameter of approximately 10 m, and run at a distance of 24 m from one another at a depth of about 25 m below the ground surface on Via dei Fori Imperiali. The distance of the axis of the closest running tunnel from the apse was about 22 m.

Figure 29 shows the stratigraphy in the direction orthogonal to the tunnel axis, and the very variable thickness of the made ground and of the medium and fine grained soils of the Palotevere. The tunnels are mainly contained into these soils, with only their bottom part lying into the layer of gravel and sand at the bottom of the Paleotevere, immediately above the base formation of the Pliocene clay.

Figure 30 summarises the main index and physical properties of the foundation soils together with the groundwater pressure distribution, which is nearly hydrostatic with a level of about 15 m a.s.l. The strength and compressibility characteristics are given in Figure 31. From the profiles in Figures 30 and 31 it is possible to recognise the Pleistocene and Pliocene formations and their different degree of overconsolidation.

7.1.3   *Green field settlements*
Figure 32 shows the contours of green-field settlements, in mm, computed using the empirical relationships described above. The contours refer to the end of construction of both running tunnels and are

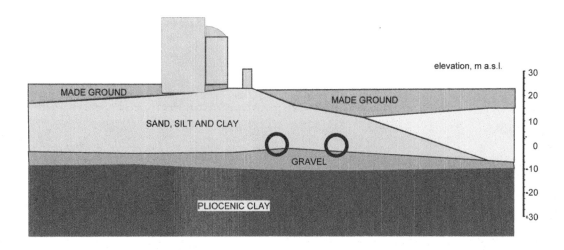

Figure 29.   Stratigraphic profile along the central transversal section.

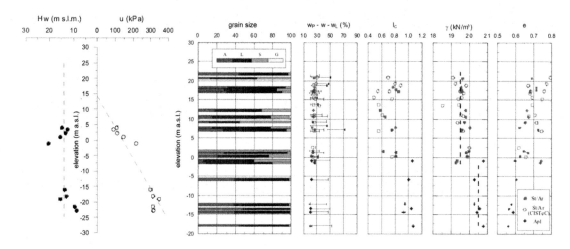

Figure 30.   Geotechnical characterisation: Groundwater pressure and main index and physical propeties.

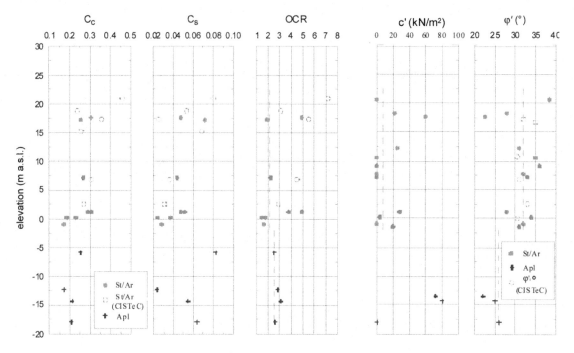

Figure 31.   Geotechnical characterisation: Strength and compressibility characteristics.

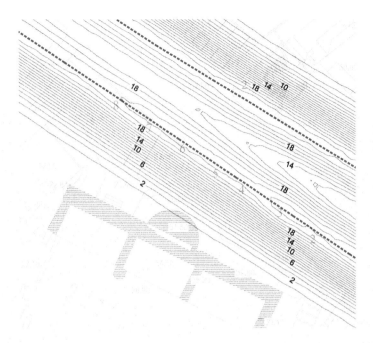

Figure 32. Contours of greenfield settlements (mm), for $V_L = 0.6\%$.

computed at the average foundation level using a volume loss $V_L = 0.6\%$. The maximum settlement occurs above the tunnel axis and is equal to about 20 mm, while approaching the Basilica the settlements progressively reduce, to values of less than 2 mm near the apse.

7.1.4 *Level 1 analysis*

The parameters required for the preliminary evaluation of potential damage were computed from the green field displacements along the perimeter of the apse, see Figure 33. For the conditions under examination, the expected damage corresponds to Category 0 of Burland's classification, that is negligible.

Following the methodological approach described above, the green field displacements were applied by the structural engineering group to both linear and non linear finite element models of the Basilica; Figure 34 and 35 show a selection of results of the structural analyses, in terms of contours of computed tensile strains in the structure.

The results obtained using the linear model indicate that the most critical conditions occur for the apse, particularly at the junction with the recent reinforced concrete dome. It is worth noting that the state of stress in the structure is mainly due to self-weight and that it is only slightly modified by the displacements induced by tunnelling. In fact, the maximum tensile strain in the structure before the passage of the tunnels is equal to 8×10^{-4}, and increases to a value just larger than 9×10^{-4} (see Figure 34); before the construction of the tunnels, according to Burland's classification, the structure already is in a state of slight damage, and the category of damage is not worsened by the works.

The maximum tensile strains obtained from the non linear analyses are generally larger than those obtained using the linear model, but, once again, these are mainly due to self weight; the maximum tensile strains in the monument after construction of the tunnels increases from 3.06×10^{-3} to 3.11×10^{-3}, or a potential severe damage (Category 4). However, the incremental tunnelling induced damage can be considered negligible.

7.1.5 *Level 2 analyses*

Level 2 analyses were carried out using models of increasing geometric complexity. The behaviour of the soils was always modelled as elasto-plastic using the Hardening Soil model; the values of mechanical parameters were obtained from the available *in situ* and laboratory tests.

121

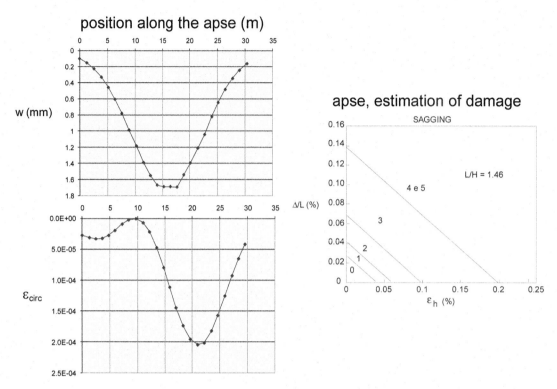

Figure 33.   Level 1 geotechnical analyses: Assessment of damage for the apse of the Basilica.

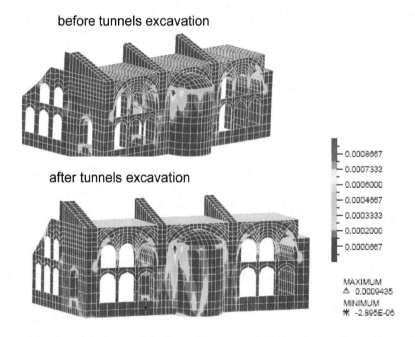

Figure 34.   Level 1structural analyses (linear model): Contours of tensile strain.

A first set of analyses was carried out in plane strain conditions, in the three sections indicated in Figure 28, orthogonal to the tunnel axes and through the piers of the Basilica. As the plane strain analyses assume that the structure is continuous in the longitudinal direction, it was necessary to scale the stiffness and weight to establish equivalence between the 3D signatures of the structure and its 2D representation.

Figure 36 shows the cross section of the Basilica through the central pier, together with the mesh used in the numerical analyses; in the same figure, a diagram illustrates the geometric assumptions adopted in the factorisation of the weights.

122

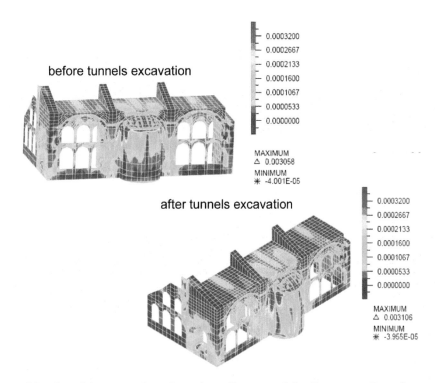

Figure 35.   Level 1structural analyses (non-linear model): Contours of tensile strain.

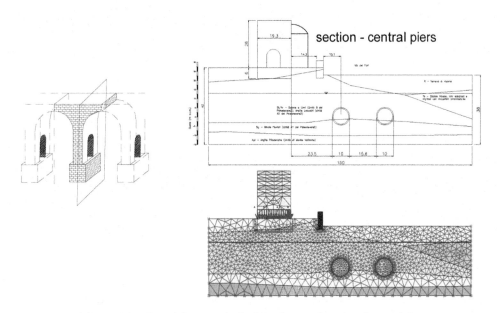

Figure 36.   Level 2 geotechnical analyses, plane strain model.

A selection of the results of the plane strain analyses are given in Figure 37; these results confirmed that the strain level in the piers is rather small, as expected from Level 1 analyses, but also provided evidence of some unexpected behaviour.

As illustrated in Figure 37, the vertical displacements of the pier reach a value of about 6 mm and are significantly larger than the corresponding settlements in green field, contradicting the common belief that green field analyses are more conservative than interaction analyses. The reason for this unusual behaviour, unique in all the studies carried out for Contracts T2 and T3, must be sought for in the structural features of the Basilica. As a matter of fact, this building has an extremely small value of the ratio of the area of the bearing structural members to the total covered area (about 12%, compared

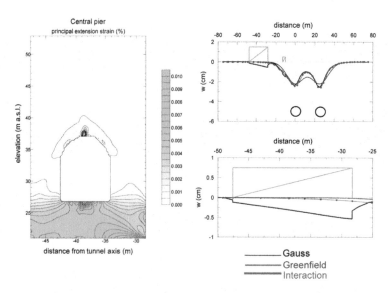

Figure 37.   Results of plane strain Level 2 geotechnical analyses.

to, *e.g.*, 23% of the Pantheon and 26% of S. Peter's Basilica); this is likely to be responsible for its high structural vulnerability and the many collapses experienced through the course of its history.

Because of the significant weight of the structure, the foundation soils experience significant deformations, considerably reducing their shear stiffness. This reduced stiffness is well reproduced by the constitutive model used in the numerical analyses, which accounts for the dependency of soil stiffness on strain level, so that even the relatively small changes of stress state at foundation level due to the excavation of the tunnels produce appreciable settlements due to the reduced local values of the shear stiffness.

Level 2 structural analyses were carried out using a more detailed three-dimensional model of the monument (see Figure 38) while additional three-dimensional soil-structure interaction analyses were performed by the geotechnical engineering group in which the structure was replaced by an equivalent solid (see Figure 39).

The results of these analyses are shown in Figure 40, with reference to the same cross-section considered above and a longitudinal section corresponding to the façade of the Basilica. The results, comparable with those obtained with the 2D analyses, confirm the relatively large settlements of the foundations of the transverse piers, with the exception of the Carinae pier; this is probably due to the fact that this pier is founded at a larger depth than the other two and induces smaller shear stress in the foundation soils that suffer to a lesser extent by shear stiffness degradation.

Figure 41 shows the results of Level 2 structural analyses in terms of contours of tensile strain, this time using a more reliable structural model in which material properties were obtained from a number of mechanical tests of samples of the masonry and non-destructive site investigations.

The results of all the analyses described above confirm the high vulnerability of the building, that must be considered as a unique case from the point of view of its structural behaviour, and demonstrate the validity of the approach taken on structural damage and its modelling, that should use non-linear models in the presence of a non negligible expected damage.

As a final comment it is worth to mention that successive modifications of design have led to a substantial reduction of the diameter of the running tunnels, with significant benefits on the potential damage to the monument, which is still receiving special attention and will require the implementation of protective measures.

7.2 *Amministrazione Doria Pamphili*

The second example that will be presented in some detail is that of the seat of the Amministrazione Doria Pamphili included in Contract T2 and located on Via del Plebiscito between Via della Gatta and Vicolo Doria, see Figure 42.

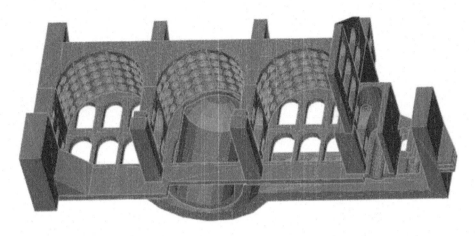

Figure 38.   Reference geometrical model for Level 2 structural analyses.

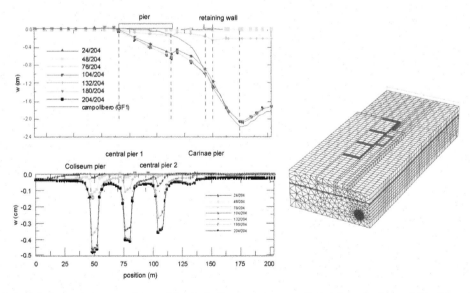

Figure 39.   The equivalent solid for Basilica di Massenzio.

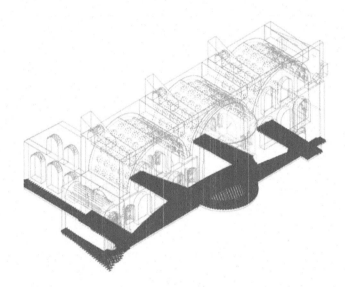

Figure 40.   Level 2 interaction analyses: Vertical displacements of the foundation of Basilica di Masse zio obtained using equivalent solid.

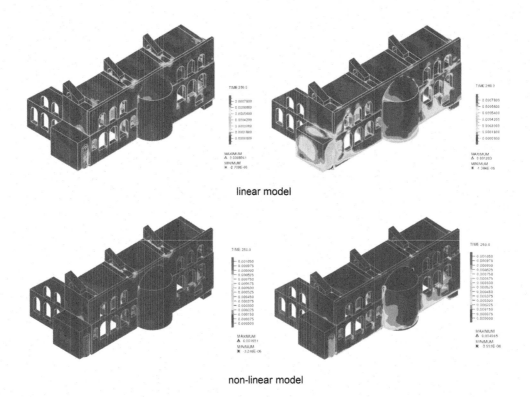

linear model

non-linear model

Figure 41. Level 2 structural analyses: Tensile strains: (a) before and (b) after tunnel excavation.

Figure 42. The building of the Amministrazione Doria Pamphilj.

This is a building consisting of four bodies constructed at different times. The main body (A) was built in the middle of the eighteenth century; body B, located in the second courtyard of the main body, and body C are both from the end of the nineteenth century; body D is the oldest in the complex, dating back to the sixteenth century, although it was raised at the end of the nineteenth century. The foundations of the building are at a depth of 6 to 7 m from the road surface.

In Piazza Venezia, coming from Via dei Fori Imperiali and before entering Via del Plebiscito, the running tunnels describe a wide curve and the first tunnel underpasses the building under examination, see Figure 43.

Figure 44 shows the green-field settlements computed using the empirical relationships of Level 1 analyses and the effect of the progressive construction of the first and second tunnel, for a volume loss $V_L = 0.5$ and 1%. After completion of both tunnels, the maximum settlements occur above the tunnel axis and are equal to about 6 mm for $V_L = 0.5\%$ and twice this value (12 mm) for $V_L = 1\%$.

126

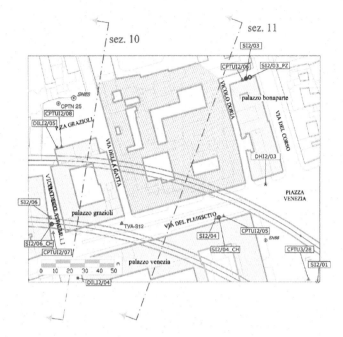

Figure 43.   Plan view of the building with the tunnel layout.

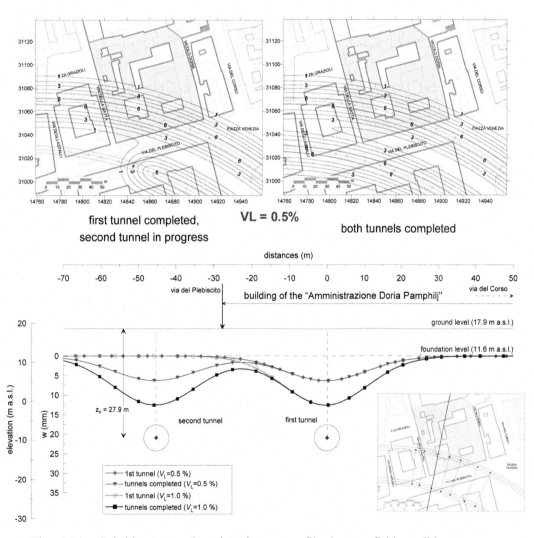

Figure 44.   Subsidence trough and settlement profiles in greenfield conditions.

127

The evaluation of the potential damage for the building, involving the computation of all the components of the displacement field also for intermediate positions of the tunnels face, led to estimate negligible damage in sagging, for both values of volume loss, and very slight damage in hogging, but only for a volume loss $V_L = 1\%$, see Figure 45.

These estimates were confirmed by the results of Level 1 structural analyses, reported in Figure 46, which emphasise, however, that the tensile strain in the walls of the foundation may reach values of $6 \div 7 \times 10^{-4}$.

The geotechnical conditions of the foundation soils, explicitly modelled in Level 2 geotechnical analyses, are the most critical along the central stretch of the line. Below the made ground, with a thickness of about 12 m, there is a layer of about 30 m of fine grained deposits, consisting of silts and clays, slightly overconsolidated in the upper part and normally consolidated in the lower part, underlain by the formation of gravel and sand immediately above the base formation of Pliocene clay, see Figure 47.

Figure 48 summarises the main physical and mechanical properties of the fine-grained soils, as obtained from the geotechnical characterisation. These are medium to high plasticity soils with low activity, with mediocre mechanical properties.

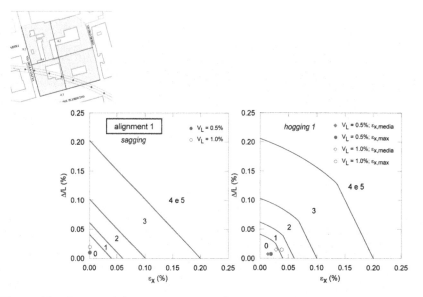

Figure 45.   Expected damage corresponding to greenfield subsidence profile.

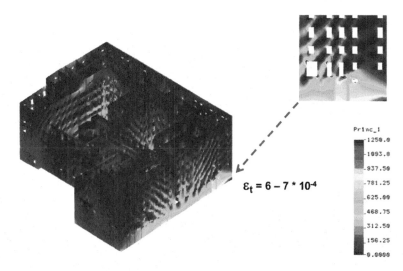

Figure 46.   Level 1structural analyses: Maximum tensile stress for 1.0% volume loss.

128

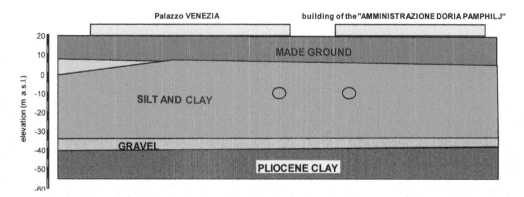

Figure 47.   Stratigraphic profile along section 11 of Figure 42.

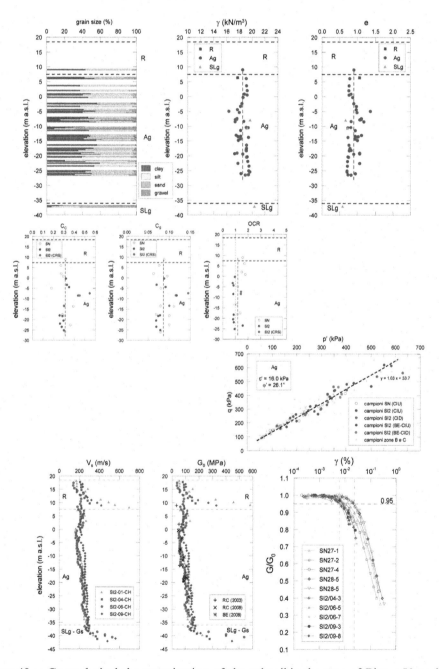

Figure 48.   Geotechnical characterisation of the subsoil in the area of Piazza Venezia.

129

The special geotechnical conditions and the high density of heavy and stiff buildings interfering with one another, suggested that three-dimensional interaction analyses should be carried out using a large model that encompassed all the relevant buildings located in the area, included in the model as equivalent solids, see Figure 49.

Figure 50 shows the results of the three-dimensional interaction analyses in terms of computed settlements and tensile strains along the most critical section (section 1). The building deforms both in hogging and in sagging, with short term settlements of the order of 10 mm, and long term settlements almost twice the short term value, due to radial seepage towards the tunnel and consolidation.

Even with relatively high values of absolute and differential settlement, the tensile strains in the structure are rather small and, for a volume loss of 0.5%, the estimated damage is negligible. Also, although long-term effects produce a significant increase of the settlements, their effect on the estimated potential damage is beneficial, as they reduce differential settlements and curvatures.

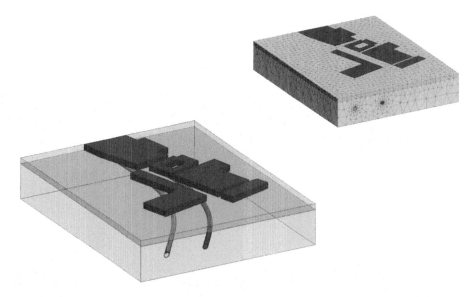

Figure 49.   Three dimensional mesh and equivalent solids used for Level 2 geotechnical interaction analyses.

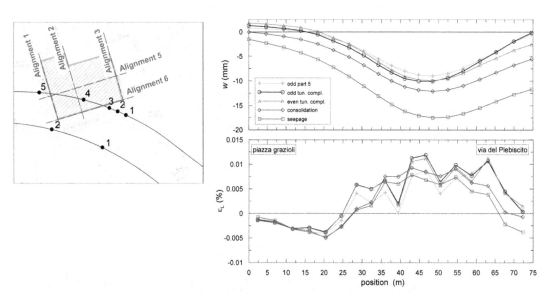

Figure 50.   Results of 3D geotechnical analyses along the most critical direction (alignment 1).

130

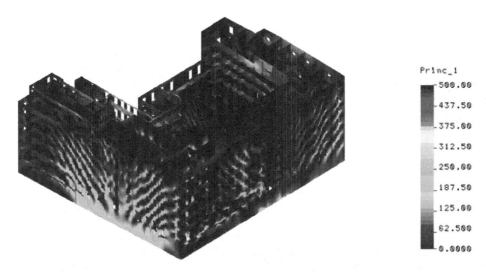

Figure 51. Results of Level 2 structural analyses.

Figure 51 shows a selection of the results of Level 2 structural analyses, confirming the conclusions obtained from the geotechnical analyses, regarding the magnitude and distribution of the tensile strain in the structural members; once again, it is found that the tensile strains are reduced in the long term.

8 MONITORING AND MITIGATION MEASURES

In a work of such importance, monitoring plays a strategic role for different objectives, such as the quantification of the background noise prior to construction, the corroboration of the design predictions, the validation of the models developed for the interaction analyses and the calibration of geotechnical parameters, the support to decision-making for the implementation of mitigation and protective measures.

All of these activities require monitoring devices targeted for the specific objective and monitoring must be scheduled to take place with different frequencies in time.

In general, geotechnical and structural monitoring require the use of standard equipment for the measurement of rotations, surface and subsurface displacements, pore pressures, and the opening of structural joints and cracks. A common feature of all these monitoring systems is the required accuracy, because of the small effects expected from the construction of the line. The robustness and durability of the instrumentation are also of importance, in view of the long observation times; for example, pore pressure measurements will have to be extended after the end of construction to monitor long-term effects.

Another factor to consider is the advancement rate of the TBM, of the order of tens of meters per day, which imposes intense monitoring sessions during the passage of the tunnel face near the monument under examination and that renders it difficult to read those instruments, such as manual inclinometers and extensometers, that require relatively long measurement times. This is why the installed instrumentation should include automatic instruments such as electric piezometers and multi-base extensometers and inclinometers.

The extension and importance of the works and the need to protect the existing structures also suggest the installation of a geomatic net of automatic total stations, see Figure 52.

These are a very effective means for the automatic measurement of the position is space of targets (retroflector prisms) located in large numbers on the façades of the buildings at distances of up to a hundred meters. The most recent and advanced total stations reach an accuracy ranging from a few tenths of a millimetre to a few millimetres; the upper bound of this range can be reduced with specific

131

Figure 52. Location of a motorized total station for the automatic measurement of the displacements of the surrounding buildings.

statistical elaboration of the data to bring the variations of the coordinates within a predetermined level of probability. This requires high level data management skills and advanced geomatics knowledge, and a specific Working Group, devoted to Geomatics, had to be set up.

As already mentioned, one of the main objectives of monitoring is to provide support to decision-making for the implementation of mitigation and protective measures. This is the classical use of monitoring under the observational approach in geotechnical design, which will be applied for the implementation of any measure intended to safeguard the monuments.

An example of mitigation measure typically carried out using the observational approach is that of compensation grouting. Before describing this technique, it may be useful to return to the criteria for the assessment of potential damage.

The current criteria for the classification of damage are based on the ease of repair. Excluding very severe damage, which clearly has irreversible consequences, the adoption of these criteria implicitly accepts that a building can always be repaired, classifying the severity of damage on the basis of the costs of the required interventions.

This pragmatic approach, commonly adopted for ordinary buildings and deriving from the balance between the costs of repair and the social benefits associated with the implementation of a new infrastructure, may not be applicable to the case at hand in which the existing structures have an exceptional historical, monumental and artistic value. In this case it is necessary to adopt more stringent criteria to define acceptable thresholds, especially for very low, almost "cosmetic", levels of damage, but also to provide appropriate mitigation measures and criteria for their conditional activation.

Compensation grouting is a technique that is being increasingly used to control ground and building movements during tunnelling in soft ground. Figure 53 illustrates the principles of the method. During tunnelling, grout is injected from tubes installed in the soil between the tunnel and the building foundations (in this specific case, below the archaeological layer) to compensate for ground loss and stress relief caused by the tunnel excavation. The sleeved grout tubes (*tubes á manchette* or TAMs) are installed in the ground prior to tunnelling, often arranged radially from vertical shafts or from the surface by directional drilling. Installation of the TAMs can cause settlements, but these can be minimised by various means. Before tunnelling commences, conditioning grouting is undertaken to tighten the ground, fill any existing voids and reverse any settlement or loosening of the ground caused by drilling for TAM installation. Conditioning is usually achieved by injecting grout in a uniform density over a

132

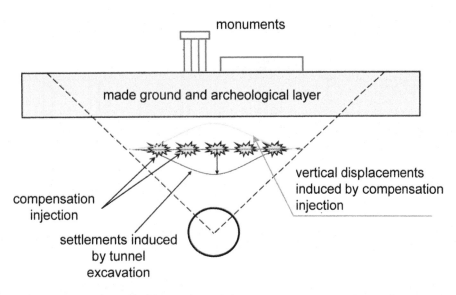

Figure 53. Schematic illustration of the compensation grouting technique.

prescribed area until instrumentation installed on the ground surface or on overlying structures begins to respond, thus showing that the ground is fully tightened. Grout injection is then undertaken simultaneously with tunnelling in response to detailed observations, the aim being to limit building settlements and distortions to specified, acceptable amounts.

The importance of site coordination and high level site supervision of compensation grouting cannot be overemphasized. Properly conducted field trials prior to tunnelling are vital to prove the feasibility of TAM installation and the proposed grout mixes, and to validate assumptions regarding grout spread.

In addition to tunnelling, the construction of the deep open excavations required to accommodate the new stations may result in ground movements that can induce damage to adjacent buildings. If the excavation is supported by cast *in situ* reinforced concrete diaphragm walls, the displacements of the ground may result both from the installation of the panels by slurry trenching and concreting, and from the actual excavation in front of the diaphragm wall. Generally, to limit the horizontal displacements and hence the induced settlements of the buildings during excavation, stiff diaphragm walls and props installed as the excavation progresses are used. However, vertical stress relief associated to excavation induces surface settlements even when the retaining walls are prevented from moving horizontally, and deep-seated inward displacements of the walls that cannot be controlled by the props that are installed within the excavation itself. Therefore, to minimise the impact of deep excavations in urban areas, inward displacements of diaphragm walls and surface settlements around excavations may be reduced installing an internal support system below the formation level, prior to excavation.

Sacrificial cross-walls may be formed by jet-grouted columns or unreinforced panels installed with diaphragm walling equipment. They are installed between the perimeter diaphragm walls before the start of excavation and are excavated out with the soil down to the depth of the final excavation level, see Figure 54. At the end of the excavation, the only portion of the cross-walls remaining to resist to the horizontal deflection of the diaphragm walls is that extending below the bottom of excavation. The need to demolish without too much difficulty sacrificial panels requires that they be designed seeking an effective compromise between their stiffness, necessary for the performance of their duty, and their strength, which should be low enough to make them easily excavated by ordinary equipment. Careful consideration is also required for the retaining wall design, to account for increased moments and shears.

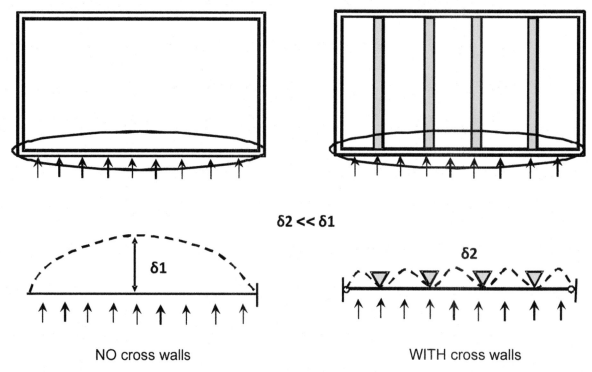

$$\delta 2 << \delta 1$$

NO cross walls   WITH cross walls

Figure 54.   Principle of operation of sacrificial cross-walls.

9   CONCLUSIONS

Contracts T2 and T3 of the new line C of Rome underground underpass the historical centre of the city where masonry buildings of particular relevance are present, mostly built between the XV and the XIX centuries. A reliable evaluation of the potential damage induced by tunnel excavation to the existing building is essential in order to proceed with design, implementing where necessary appropriate mitigation techniques.

The procedure developed for evaluating these effects hinged on the geotechnical analyses, starting from a careful geotechnical characterisation based on *in situ* and laboratory tests and including the use of reliable computation models, but at the same time promoted a fruitful interaction of the geotechnical and structural engineers. At several stages, parallel evaluation of the damage to the buildings were carried out by both groups using the tunnelling-induced displacement fields computed in the geotechnical analyses. These independent estimates by the geotechnical and the structural engineers always provided consistent results.

Evaluation of tunnelling-induced effects was carried out following procedures of increasing level of complexity. Level 1 green-field evaluations were carried out using empirical relationships and assuming that the buildings follow the ground displacements; in some cases, at a subsequent stage, depending on the results of Level 1 analyses and on the specific relevance of the building, Level 2 interaction analyses were carried out in which the influence of the weight and the stiffness of the building was explicitly considered using a simplified description of the building through an equivalent solid entirely embedded into the soil, down to the foundation level. In addition to the study of the soil-structure interaction, these numerical analyses permitted to evaluate the long-term settlements that may develop when tunnels are excavated in fine-grained soils of very low permeability.

As a general result, explicit consideration of stiffness and weight of the building resulted in somewhat larger settlements but smaller distortions, and therefore predicted a lower damage if compared with the green-field. The reduction in the curvature of long-term settlement troughs resulted in a reduction of the predicted damage to the buildings.

In the evaluation of the potential damage it is very important to assess the magnitude of the background noise, *i.e.* of the movements of buildings due to external causes, such as daily and seasonal temperature cycles, fluctuations in the water level of the Tiber river, as it provides the reference lower bound of acceptable movements.

Finally, accurate monitoring and provision of measures for the protection of the monuments form the effects of the construction are two key aspects of the whole process that can be effectively implemented only if they are included consistently as part of the observational method in geotechnical design.

ACKNOWLEDGMENTS

The Authors are grateful to the Grantor, Roma Metropolitane SrL, and the General Contractor, Metro C SpA, for making this unique geotechnical experience possible. Special thanks are also due to the other members of the Technical and Scientific Steering Committee, and particularly to profs. Jamiolkowski, Carbonara, Kovari, and Macchi. The Authors also remember fondly the late prof. Funiciello, who died in 2009, leaving a huge void in the Committee.

REFERENCES

Boscardin, M.D. & Cording, E.J. (1989). Building response to excavation—induced settlement, *Journal of Geotechnical Engineering, ASCE*, vol. 115(1), 1–21.

Burland, J.B. (1995). Assessment of risk of damage to buildings due to tunneling and excavation, *1st Int. Conf. on Earthquake Geotechnical Engineering*, Tokyo, 1189–1201.

Burland, J.B., Broms, B.B. & de Mello, V.F.B. (1977). Behaviour of foundations and structures—SOA report, *9th Int. Conf. on Soil Mechanics and Foundations Engineering*, Tokyo, vol. 2, 495–546.

Burland, J.B. & Wroth, C.P. (1974). Settlement of buildings and associated damage, *Proc. Conf. Settlement of Structures*, Cambridge, UK, 611–654.

Calabresi, G., Cassinis, C.e. & Nisio, P. (1980)—Influenza del regime del Tevere sul comportamento di un fabbricato monumentale adiacente. *XIV Convegno Nazionale di Geotecnica*. Firenze.

Finno, R.J., Voss jr. F.T., Rossow, E. & Blackburn, J.T. (2005)—Evaluating Damage Potential in Building Affected by Excavations. *Journal of Geotechnical and Geoenvironmental Engineering, ASCE*, Vol. 131, No 10, October, 1119–2100.

Mair R.J., Taylor R.N. & Bracegirdle A. (1993). Subsurface settlement profiles above tunnels in clay, *Geotechnique*, vol. 43(2), 315–320.

Mair, R.J. & Taylor, R.N. (1997)—Bored tunneling in the urban environment. State-of-the-art report and theme lecture. *Proc. 14th Int. Conf. Soil Mech. Found. Engng*, Hambourg **4**, 2353–2385.

Mair, R.J. (2008)—46th Rankine Lecure. Tunnelling and geotechnics: new horizons. *Géotechnique* 58, No 9, pp. 695–736.

Mazars, J. (1984)—Application de la mécanique de l'endommagement au comportement non linéaire et à la rapture du béton de structure. *Thèse de Doctorat d'Etat*, L.M.T., Université Paris, France.

Moh, Z.C., Hwang, R.N. & Ju, D.H. (1996). Ground movements around tunnels in soft ground, *Proc. Int. Symp. on Geotechnical Aspects of Underground Construction in Soft Ground*, London, 725–730.

O'Reilly, M.P. & New, B.M. (1982). Settlements above tunnels in the United Kingdom—Their magnitudes and prediction, *Proc. Tunnelling '82 Symp.*, London, 173–181.

Peck, R.B. (1969)—Deep excavations and tunnelling in soft ground. State of the art report, Mexico City, *State of the Art Volume. Proc. 7thInt. Conf. SMFE*, pp. 225–290.

Pickhaver, J.A. (2006)—Numerical Modelling of Building Response to Tunnelling. *PhD Thesis*. University of Oxford.

Rankin, W.J. (1988)—Ground movements resulting from urban tunneling; predictions and effects. Engineering Geology of Underground Movement, Geological Society. *Engineering Geology Special Pubblication* No 5, pp. 79–92.

Ribacchi, R. (1993). Recenti orientamenti nella progettazione statica delle gallerie, *XVIII Convegno Nazionale di Geotecnica*, Rimini, vol. 2, 37–92.

Schanz, T., Vermeer, P.A. & Bonnier, P.G. (1999). Formulation and verification of the Hardening—Soil Model. *Proceedings Plaxis Symposium "Beyond 2000 in Computational Geotechnics"*, Amsterdam, 18–20 March 1999, 281–296.

Schmidt, B. (1969). Settlements and round movements associated with tunnelling in soil, *Ph.D. Thesis*, University of Illinois, p 224.

Voss, F. (2003)—Evaluating Damage Potential in Building Affected by Excavations. *MSc thesis*, Northwestern University Evanston, IL. p 166.

Wongsaroj, J. (2005)—Three-dimensional finite elements analysis of short—and long-term ground response to open face tunneling in stiff clay. *PhD Thesis*, Cambridge University.

Geotechnics and Heritage – Bilotta, Flora, Lirer & Viggiani (eds)
© 2013 Taylor & Francis Group, London, ISBN 978-1-138-00054-4

Protective compensation grouting operations beneath the Big Ben Clock Tower

D.I. Harris
DI Harris Geotechnics Ltd., UK

R.J. Mair
University of Cambridge, Cambridge, UK

J.B. Burland & J.R. Standing
Imperial College of London, London, UK

ABSTRACT: The construction of the Jubilee line Extension Station at Westminster, London was predicted to produce significant movements of the Big Ben Clock tower and the adjoining Palace of Westminster. The works consisted of the excavation of two 7.4 m tunnels and a 39 m deep station escalator box located 34 m north of the foundations of the Clock Tower. The protective measures adopted to minimise damage to these priceless historic buildings consisted primarily of compensation grouting below the Clock Tower and proved extremely effective in controlling settlement and tilt of the structure. This case study not only demonstrates the success of this innovative protective measure, but also shows the value of careful interpretation of appropriate numerical modelling and the results of high quality monitoring during and after completion of the works.

1 INTRODUCTION

The Big Ben Clock Tower was constructed in 1858, and consists of load-bearing brickwork with stone cladding approximately 11 m square to a height of 61 m, supporting a cast iron framed belfry and spire to a total height of 92 m. The Tower is founded on a mass concrete raft, 15 m square and 3 m thick within Terrace Gravels overlying London Clay. The weight of the Tower is about 8400 t giving an average foundation bearing pressure of approximately 400 kPa. It is worth noting that Big Ben is the name of the largest bell in the belfry—the clock tower itself is named St Stephen's Tower.

The Clock Tower is structurally connected to the four-storey East Wing of the Palace of Westminster, as shown in Figure 1, which houses the offices of the Ministers of State. Both the Clock Tower and the Palace have a single-level basement of vaulted brickwork. The main structure of the Palace comprises load-bearing masonry with the floors formed of cast iron beams with brick jack arches spanning between them. Several masonry arch structures are included within the ground-floor structure to provide access from New Palace Yard to Speaker's Green and Speaker's Court. Like the Clock Tower, the Palace of Westminster is founded on a mass concrete raft.

The construction of the new Westminster Station on London Underground Limited's Jubilee Line Extension (JLE) project was predicted to cause significant movements of the Clock Tower and the adjoining Palace of Westminster. The station consists of two station platform tunnels, one vertically above the other, and a 39 m deep station escalator 'box' for access purposes—see Figure 1. Protective measures, primarily in the form of compensation grouting beneath the Clock Tower, were implemented during the construction period to control the settlement and tilt of the monument. This paper describes these protective measures and presents the results of the monitoring during and subsequent to the works. Figure 2 is a photograph of the Palace of Westminster taken during the construction of the station.

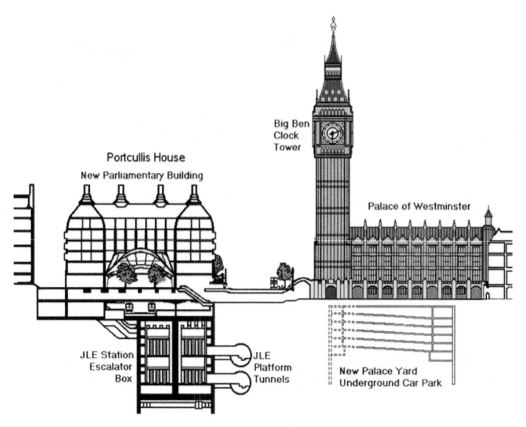

Figure 1.　Cross-Section showing Westminster JLE Station, Big Ben Clock Tower and the Palace of Westminster.

Figure 2.　View of Palace of Westminster during construction of Westminster Station.

2　THE NEW WESTMINSTER STATION AND ITS CONSTRUCTION

The layout of the new Westminster Station is shown in plan in Figure 3 and in section in Figure 4. The JLE platforms are contained within 7.4 m OD (outside diameter) bored tunnels in a vertically stacked arrangement below Bridge Street. The axes of the lower westbound tunnel and upper eastbound tunnel

138

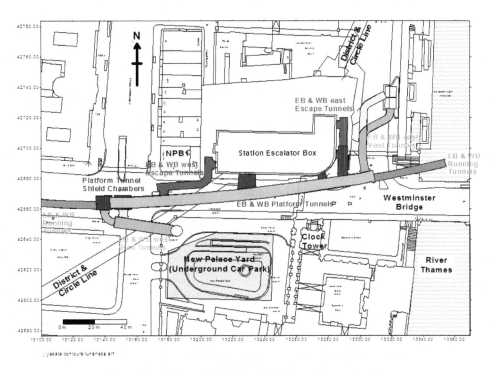

Figure 3. Plan of the new JLE Westminster Station.

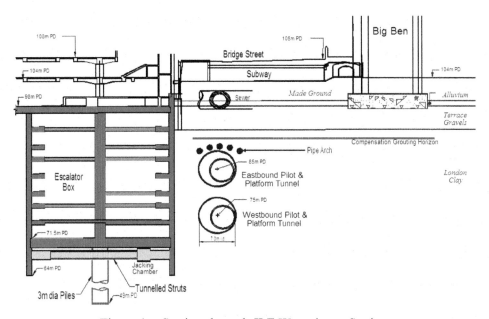

Figure 4. Section through JLE Westminster Station.

are at depths of 30 m and 21 m below ground level respectively. Access to the platforms is by means of four adits connected to the 39 m deep station escalator box to the north constructed using diaphragm walls. The escalators rise to the District and Circle Line platforms, the ticket hall and Bridge Street. The northern edge of the Clock Tower is 28 m from the centre-line of the tunnels and 34 m from the diaphragm walls of the deep escalator box. Details of the layout, the design and the construction of the station are given by Carter *et al* (1996) and Bailey *et al* (1999).

The station escalator box is 74 m by 28 m in plan and, with excavation up to 39 m below street level, was at that time substantially the deepest basement ever to have been constructed in London. The retaining walls consist of reinforced concrete diaphragm walls constructed to a maximum depth of

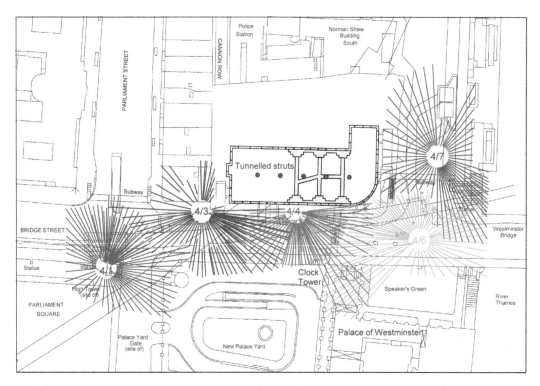

Figure 5.   Compensation grouting arrays and tunnelled struts.

40 m from a platform 4 m below ground level. The excavation was carried out using the well known 'top-down' technique with the struts and floors being installed progressively from the top downwards as excavation progressed. In order to minimise surrounding ground movements, low-level struts were installed in tunnels close to the base of the diaphragm wall prior to excavation below the main roof slab (Crawley and Stones, 1996). The system had previously been used successfully during the construction of the Barbican Arts Centre in London (Stevens *et al*, 1977). The location of the tunnelled struts is shown in elevation in Figure 4 and in plan in Figure 5.

Prior to any substantial excavation within the station escalator box, the running tunnels were driven from east to west and acted as pilot tunnels for the station tunnels. These running tunnels are 4.85 m OD and were built in expanded concrete segmental linings. A Howden open-face shield with a back actor was used for both tunnel drives. The platform tunnels were driven from 9.4 m OD shield chambers located at the west end as shown in Figure 3. As for the running tunnels, an open-face shield was also used for the enlargement to form the 7.4 m OD platform tunnels. The tunnels are 160 m long and are lined with bolted segmental SGI (Spheroidal Graphite Iron) rings. The ventilation and escape tunnels shown in Figure 3 were too far away to have any influence on the Clock Tower.

3   GROUND AND GROUNDWATER CONDITIONS

The stratigraphy at Westminster is as follows. Made Ground of depth varying between 5 m and 8 m overlies Alluvium and Terrace Gravels. The combined thickness of the Alluvium and Terrace Gravel is about 5 m. The London Clay at this location is 35 m thick and a detailed description of it is given by Burland and Hancock (1977). The Lambeth Group below the London Clay lies over Thanet Beds which are above the Chalk. The Lambeth Group is 18 m thick. It is predominantly clayey comprising 8 m of Upper Mottled Clay dissected by a thin layer of Laminated Beds, over Lower Mottled Clay about 5 m thick. The lowermost 5 m includes a thin layer of the Pebble Bed over Glauconitic Sand. The Thanet Beds are about 8 m thick on the top of the Chalk, some 73 m below ground level.

The groundwater level in the Terrace Gravels aquifer is about 9 m below ground level with little tidal variation. Pore water pressures are close to hydrostatic equilibrium with the overlying aquifer throughout the London Clay and within the Upper Mottled Clay of the Lambeth Group. The water pressure in the Thanet Sands and Chalk is substantially reduced because of historical pumping from this aquifer.

4 ASSESSMENTS OF GROUND MOVEMENTS AND POTENTIAL DAMAGE

Assessments of the differential ground movements and their impacts on the tilt of the Tower and the potential damage to the adjoining Palace of Westminster formed a vital part of the design process of the JLE Westminster Station.

The short-term ground movements associated with the tunnelling were assessed using the widely used empirical approach summarised by Attewell *et al* (1986) and Rankin (1988) in which the settlement troughs are assumed to be Gaussian normal distribution curves. A trough width factor of 0.5 was assumed for all tunnels together with volume loss values of 2 per cent for the pilot tunnels and 3 per cent for their subsequent enlargement to form the platform tunnels. The tunnel alignment was designed to maximise the plan distance from the Clock Tower. Figure 6 shows the estimated settlement troughs for each of the individual tunnels together with the resultant tunnelling settlement trough obtained by adding the individual ones (i.e. assuming superposition to be valid). The calculated short-term settlement at the northern edge of the Clock Tower was 4.5 mm with no movement at the southern edge giving a differential settlement across the Tower of 4.5 mm. It can be seen from Figure 6 that most of the tunnelling-induced settlement is associated with the construction of the deeper westbound running and subsequent station tunnel.

Finite element analyses were used to assess the short-term movements associated with the escalator box excavation and, in particular, to assess the effectiveness of measures designed to reduce these movements. Of these, the most important was the inclusion of low-level struts across the box installed prior to excavation as described previously. The restraint to movement of the walls was shown to reduce the magnitude of the induced movements of the Clock Tower by up to 40 per cent and the associated differential settlements by 30 per cent. Predictions of movement were then made by applying reduction factors based on these FE results to the empirical method of Clough and O'Rourke (1990). The predicted settlement at the north and south sides of the Clock Tower due to the excavation of the escalator box were 12.7 mm and 9.8 mm respectively, giving a differential settlement of 2.9 mm across the Clock Tower.

Superposition of the estimated movements for the tunnels and the station box indicated that a maximum short-term settlement of 21.5 mm and an increase in tilt of approximately 1:2000 could be anticipated. The greater part of the settlement was expected to result from the box excavation, whereas the tunnelling was expected to produce most of the tilt of the Clock Tower.

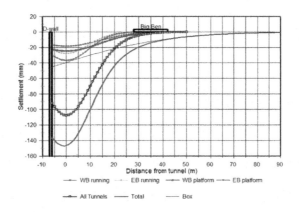

Figure 6.   Estimated settlement profiles from the tunnels and the escalator box.

141

Assessments of potential damage to the Palace of Westminster in the vicinity of the structural connection to the Clock Tower using the limiting tensile strain approach (Burland, 1995) indicated that 1:2000 was close to the maximum acceptable limit of tilt for the Clock Tower. Moreover analyses had indicated that movements could be substantially increased by interaction effects and that long-term settlements could be expected following tunnelling. It was therefore evident that additional, contingency protective measures would be needed.

5 COMPENSATION GROUTING ARRANGEMENT

Potential damage assessments were undertaken for all of the structures within the zone of ground movement adjacent to the proposed Westminster Station. These assessments indicated that protective measures would be required for several of them. Compensation grouting was specified as the way to protect them. The resulting proposal was to use vertical shafts to install horizontal grouting arrays above the crown levels of all the tunnels around the Westminster Station complex and the resulting arrays are shown in Figure 5.

Compensation grouting comprises the controlled injection of grout between the tunnel and the building foundations in response to observations of ground and building movements during tunnelling or excavation. As its name implies, the purpose is to compensate for ground loss. The technique requires detailed instrumentation to monitor the movements of the ground and the building. Harris (2001a) gives useful brief descriptions of various forms of grouting together with a detailed description of the one used on the Jubilee Line Extension involving fracture grouting. For this, liquid grout was injected into the London Clay from sub-horizontal steel tubes with ports at regular intervals. The ports comprise four holes spaced equally around the circumference of the tube and covered with a rubber sleeve (*manchette*). The grout is injected by inserting a probe into the tube and isolating the port to be injected by inflating packers at either side of the injection nozzle. Sufficient pressure is then applied to open the port and initiate flow into the ground. The tubes and ports are usually referred to as *tubes à manchettes* (TAMS).

Compensation grouting was adopted as the most appropriate method of ensuring that the tilt of the Clock Tower did not exceed the maximum specified value of 1:2000. The grouting arrays were extended below the full footprint of the Tower's foundations as shown in Figure 5. The arrays installed below the Tower initially comprised six TAMS from Shaft 4/4 with a maximum spacing of 5 m. Subsequently, as a consequence of changes in other construction activities, these arrays had to be replaced; 16 TAMS were then installed from Shaft 4/6, with maximum spacing reduced to 2.5 m. The elevation of the TAMs was constrained to be between the upper platform tunnel and the interface between the London Clay and the overlying water bearing gravels. The clay cover to the tunnel was about 5 m and the selected grouting horizon was 3 m above the crown of the upper station tunnel which was judged to give adequate cover to avoid intercepting the overlying water bearing gravels—see the vertical section in Figure 4.

6 COMPENSATION GROUTING CONTROL

The management system developed for compensation grouting was that each injection had to be defined in a grouting proposal. All injections were prescribed in terms of shaft, TAM and port number. The volume of each injection was pre-determined, and sequences of injections were specified if deemed necessary. The grouting was controlled through a monitoring control office that was also in direct communication with the tunnellers and surveyors, and had access to the real-time monitoring. Each grouting proposal incorporated communication procedures that required a positive confirmation to be received from the monitoring control office after completion of a specified part of the proposal before work could proceed. For example, with an advancing tunnel, a pattern of injections relative to the face was defined (Harris *et al*, 1996). Injection could only start once a given stage of the tunnelling cycle had been completed. Moreover the injections would have to be completed before tunnelling could proceed beyond a specific point.

To maintain the flexibility needed to modify the grouting proposals in response to observed behaviour, frequent meetings were held to review construction progress, grouting records and monitoring results. Minor modifications were made by omitting injections or changing grout volumes. If significant changes were necessary, a revised proposal would be produced. The short time-scales for production, discussion, amendment and consenting to grouting proposals, required a co-operative approach from all parties to avoid delays to the works. Further details of the strategy adopted for implementing the compensation grouting are given by Harris *et al* (2000) and Harris (2001b).

7 INSTRUMENTATION

Successful implementation of the compensation grouting technique requires reliable and accurate monitoring, its rapid processing and dissemination and informed interpretation in conjunction with records of construction activities. The following instrumentation was installed on the Big Ben Clock Tower and the adjacent Palace of Westminster.

7.1 *Tilt monitoring*

Tilt of the Clock Tower was identified as the most important parameter to monitor and consequently a range of independent systems were used.

An optical plumb had been used to monitor the tilt of the Clock Tower during the construction of the New Palace Yard underground car park in the early 1970s (Burland and Hancock, 1977) and had been read intermittently over the intervening period. The original target which was removable was still available and it was decided that the JLE surveyors should monitor this using a Wild ZL optical plumb. A new datum was established which was related to the original datum giving a self-consistent data set extending over a period of nearly thirty years. A re-designed target was procured to improve the repeatability of the readings. Observations were taken on each of four faces in the north-south and east-west directions and then averaged to give a reading. The observations were recorded to a resolution of 0.1 mm and the four independent observations generally lay within a range of 0.5 mm. The resolution and precision over the 55.4 m gauge length are equivalent to tilts of about 1:550,000 and 1:110,000 respectively.

Retro reflective prism targets attached to the north, east and west clock faces were surveyed to give displacements in three dimensions using a Leica TC1610 Total Station. The observation of each target required a separate set-up location and hence the three measurements were entirely independent. Readings were taken to a resolution of 0.1 mm for distance and 1" of arc on horizontal and vertical angles. The repeatability of the readings was ±2 mm or 1:13,000.

In order to avoid the need for excessive surveying resources during grouting episodes, a real-time monitoring system was necessary. Eight electrolevels were installed on 1 m long beams, six of which were mounted horizontally and the other two vertically. Four were oriented to measure north-south tilt and four to measure east-west tilt. In the event, these instruments were not used to control the works due to the success of an alternative real-time system developed by the contractor—an electronically monitored plumb line.

The electronically monitored plumb line was developed by the contractor's surveying department and was named the "Gedometer" after its primary creator Gerald "Ged" Selwood. The instrument comprised an invar strip suspended from a grillage in the belfry over a ventilation shaft which extends over the full height of the Tower in its north-east corner. A temporary decking was installed in this shaft 5 m above ground level (which gave an almost identical gauge length as the optical plumb) on which a digitising tablet was installed. A puck was suspended from the end of the plumb line and the tablet programmed to automatically record the location of the puck at 1 second

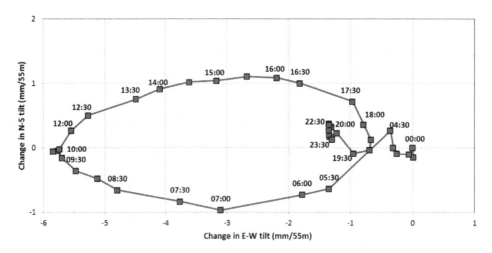

Figure 7. Measured horizontal movements at the height of the clock face taken at half-hourly intervals during a single day (18th July, 1996).

intervals. Individual observations were averaged to produce a reading at specified intervals—generally 30 minutes was found to be adequate. The instrument performed reliably over a period of 4½ years with only occasional adjustment and maintenance. In these instances corrections to the recorded movements have been necessary. The readings were reported to 0.001 mm and the accuracy was arguably as good as 0.1 mm or 1:550,000 over the period of individual compensation grouting episodes.

As part of the settlement monitoring described later, four levelling points on the corners of the Tower were used to calculate the tilt of the Tower. The accuracy of the levelling of these four points was generally ±0.3 mm and their spacing about 12 m. This gives an accuracy in the measurement of tilt of about 1:20,000.

It is important to appreciate that a structure such as the Clock Tower undergoes significant movements due to daily and seasonal temperature changes. Burland and Hancock (1977) noted that there is a seasonal east-west movement of about 6 mm at a height of 55 m with the Clock Tower moving westwards in the summer and eastwards in the winter. The measurements made with the "Gedometer" recorded a maximum daily range of tilt of 6.2 mm in the east-west direction compared to a cycle of about 2.5 mm in the north-south direction. The magnitude of the thermal bending is related to the number of hours of sunshine during the day. Consequently it is at its greatest in the summer months, when the top of the Tower traces an approximately elliptical path with its major axis in the east-west direction as shown in Figure 7. Burland and Viggiani (1994) report very similar daily movements of the Pisa Tower. An important consequence of this seasonal and daily behaviour is that it is difficult to determine reliably small long-term trends in changes of inclination of the Clock Tower. This important topic is discussed later.

7.2 *Settlement monitoring*

Settlements were monitored using precision levelling but with readings taken by three independent teams of surveyors (the contractor, the JLE Project team and Imperial College). BRE levelling sockets which had been installed to monitor movements during construction of the underground car park were augmented by additional points to provide a comprehensive system of 42 points on the Tower and the adjoining Palace of Westminster—see Figure 8. Points were also installed in the subway below Bridge Street to give a continuous settlement profile over a distance of up to 90 m from the escalator box diaphragm walls southwards. The location of the datums used by the three teams of surveyors differed but if the results are compared on a consistent basis discrepancies rarely exceeded 0.5 mm.

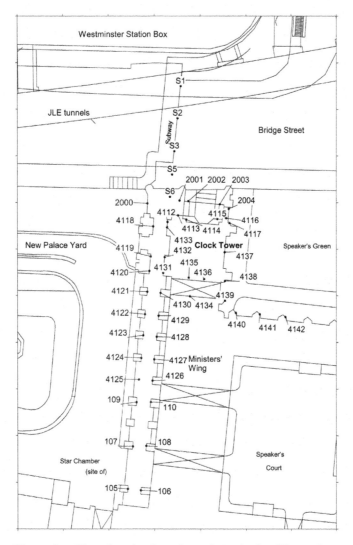

Figure 8.   Plan showing location of precise levelling points.

7.3   *Horizontal displacements*

The potential for concentration of horizontal strain on pre-existing lines of weakness within the Palace of Westminster was of particular concern. Two systems of monitoring horizontal displacements were established. A string of retro targets for total station monitoring was used by the contractor and tape extensometer measurements were taken between the precise levelling BRE sockets by the Imperial College team.

7.4   *Crack widths*

A number of pre-existing cracks identified in the pre-construction defect survey were instrumented. Demec studs in pairs, and sets of three, were installed and monitored by the contractor using a digital vernier and by Imperial College using a standard Demec gauge.

7.5   *Temperature*

Four temperature sensors were installed in the Clock Tower.

145

8 MEASURED TILT OF THE CLOCK TOWER DURING CONSTRUCTION

Monitoring of the Clock Tower for tilt began in November 1994. Monitoring was carried out throughout construction and continued for 15 years after construction ended. Figure 9 shows the measured North-South tilts of the Clock Tower from the optical plumb throughout the construction period and for six months thereafter. The timings of the main construction activities are also shown in the figure. The passage of the four tunnel drives are shown across the top of the figure and the timing at various excavation depths in the escalator box are shown across the bottom. The thick vertical line at December 1995 indicates the start of compensation grouting to control the tilt of the Clock Tower and the various episodes of grouting are shown across the middle of the figure, the final episode being in September 1997.

Initially performance control levels (PCL) of tilt were set conservatively at 1:6,000 and 1:4,000 for the amber and red triggers. These limits are equivalent to about 9 mm and 14 mm relative northward movement at the height of the clock face.

As predicted, the first activity to affect the tilt of the Clock Tower was the westbound running tunnel drive, which was undertaken in March 1995 with limited compensation grouting. A tilt to the north of about 4 mm was observed as shown in Figure 9. The northward tilt continued to increase significantly following the tunnel drive and reached the amber PCL of 9 mm in July 1995. This initiated a reappraisal of the potential movements. It was also evident that it was necessary to minimise further movements and to develop a strategy for implementing compensation grouting.

Back-analysis of the observed settlements associated with the westbound tunnel indicated that the volume losses in this area were about 3 per cent—significantly larger than had been allowed for in the design settlement assessments. The reasons for this are discussed by Standing and Burland (2006). The observed increase in tilt after the westbound tunnel drive also demonstrated that substantial time-dependent movements should be expected both during the construction period and subsequently. The following actions, *inter alia*, were undertaken.

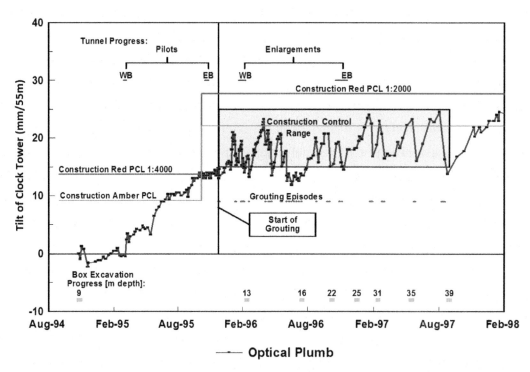

Figure 9.   Optical plumb measurements of the Northward tilt of the Clock Tower during the works.

146

1. Co-ordination of tunnel advance with implementation of grouting to allow settlements to be fully compensated; this was referred to as *concurrent* compensation grouting.
2. The red PCL on the permissible increase in the northward tilt of the Clock Tower was raised to 1:2,000.
3. The amber PCL on the tilt of the Clock Tower at which grouting would be instigated was raised to 1:2,500.
4. A trial grouting episode below the Clock Tower was undertaken to demonstrate that control of the tilt of the Tower could be exercised.
5. An expert review panel was set up to advise on geotechnical and construction issues relating to the Clock Tower.
6. Close liaison with the Parliamentary Works Directorate was to be maintained through its geotechnical advisor.
7. The finite element analysis undertaken at the design stage was updated to take account of revised construction methods and sequences. The analysis was calibrated against the observed settlements to give the best possible prediction of future movements. The results of this finite element analysis assisted in identifying potential mechanisms of movement of the Clock Tower and the adjoining Palace of Westminster and allowed variations in the excavation and construction procedure of the station escalator box to be investigated.
8. Further independent measurements of horizontal and vertical movements of the Palace of Westminster were made by Imperial College as part of a major research project on the JLE (Burland *et al*, 1996).

The shallower eastbound tunnel drive was permitted to go ahead in October 1995 with concurrent compensation grouting, before the trial grouting below the Clock Tower mentioned in item 4 above, because the Clock Tower was outside its zone of influence. As can be seen from Figure 9 no noticeable increase in tilt took place at this time.

The trial grouting was carried out in December 1995 at which time the northward tilt was 14 mm. The trial was inconclusive and a further trial was delayed until February 1996 due the necessity of installing new TAMs below the Clock Tower from Shaft 4/6—see Figure 5. During the second trial the tilt was reduced by about 5 mm confirming both the suitability of the method and that significant control could be exercised.

The next grouting episode below the Clock Tower was concurrent grouting associated with the enlargement to form the westbound station tunnel. Grouting within the settlement trough was fully coordinated with tunnel advances and was augmented by additional injections below the Clock Tower. The aim was to produce full compensation for the tunnelling-induced settlements together with a

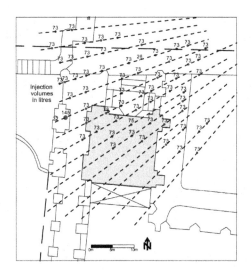

Figure 10.   Typical grout injection pattern.

small reduction in tilt. This was successfully achieved, with the tilt being reduced by 5 mm. Thereafter, during excavation of the escalator box from 9 m depth to 39 m and the enlargement for the eastbound platform tunnel, grouting was undertaken in response to the observed tilts of the Clock Tower rather than being directly related to construction activities. A construction control range of 15 mm to 25 mm tilt was adopted. It can be seen from Figure 9 that the upper limit of this control range was not exceeded throughout the construction period although occasionally the lower limit was exceeded. In total, 24 episodes of grouting were undertaken between January 1996 and September 1997 over which period a total volume of 122 m³ of grout were injected beneath the Clock Tower. In general, grouting was confined to the northern half of the raft foundation as shown in Figure 10 for a typical grouting episode. Harris (2001b) gives a more detailed account of the grouting procedures adopted.

9 SETTLEMENT MONITORING DURING CONSTRUCTION

The locations of the monitoring points are shown in Figure 8 and point106 was usually taken as the datum. Figure 11 shows the vertical movements with time for a selection of the points along the western façade of the main structure of the Palace of Westminster. Points 4112 and 4131 are located on the Clock Tower itself. The timing of the main construction events and the compensation grouting episodes below the Clock Tower are indicated on Figure 11 and the observed vertical displacements clearly show the cycles of heave associated with the grouting episodes and the subsequent settlement. The following additional features should be noted:

1. The settlement at the start of grouting in December 1995 had a maximum value of 9 mm at point 4112 on the north-west corner of the Clock Tower. This had increased to 11 mm at the time of the second successful trial in February 1996.
2. During the 20 month period of grouting to control the tilt of the Clock Tower (February 1996 to September 1997) the settlement of point 4112 was in the range 7–14 mm. In other words, during the period of grouting the vertical movement was controlled between 4 mm heave and 3 mm settlement.
3. Immediately before the final grouting episode on 28th August 1997 the maximum settlement of point 4112 had reached 14 mm and after completion of this episode it had reduced to less than 9 mm. i.e. a heave of about 5 mm was induced.

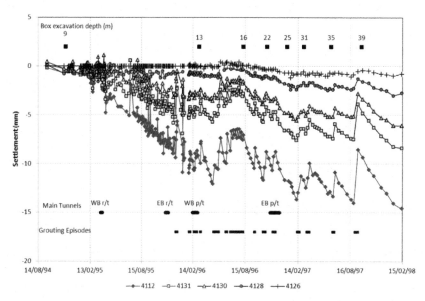

Figure 11. Vertical movements for a selection of the points on the Palace of Westminster during compensation grouting (see Fig. 8 for location of leveling points).

148

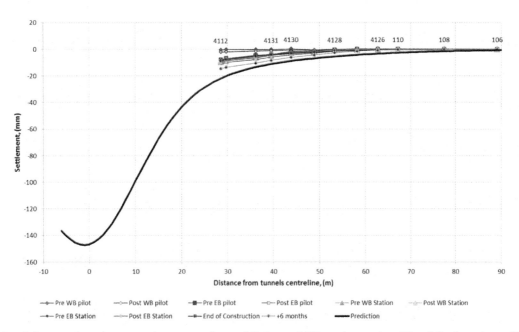

Figure 12.   Measured settlements along west face of Palace of Westminster (see Fig. 8 for location of levelling points).

4. In the five months following the final grouting episode the settlement increased to about 15 mm thereby returning approximately to the value before the final grouting episode.
5. The maximum heaves associated with individual grouting episodes were generally 3 mm or less with the exception of the final episode where up to 5 mm heave was produced.

Figure 12 shows the measured settlements along the west façade of the Palace of Westminster at various stages. Note that points 4112 southwards to 106 are located on the Palace of Westminster as shown in Figure 8. The measurements north of point 4112 relate to points located on minor structures abutting the Palace of Westminster and the Bridge Street pedestrian subway and are not included. Also shown in Figure 12, for the purposes of comparison, is the resultant estimated settlement trough with no compensation grouting taken from Figure 6.

The settlement observations plotted in Figure 12 correspond to immediately before and immediately after each of the four main tunnel drives and from the end of construction at the September 1997.

The following observations relate to the settlements of the western façade of the Palace:

1. Settlements extend southwards to a distance of 30 m from the north face of the Clock Tower (i.e. from survey point 4112).
2. The shape of the settlement profile is similar to the resultant estimated settlement trough but the magnitude is substantially less.
3. Even though the compensation grouting only extended beneath the northern half of the Clock Tower its effects extended southwards as far as point 4127 which is about 25 m from the plan extent of the grout injections.

10   MEASUREMENTS OF HORIZONTAL STRAIN AND CRACK WIDTHS

In addition to the settlement monitoring of the Palace of Westminster, horizontal displacement measurements were made using a tape extensometer and crack-width measurements were taken across significant pre-existing cracks. The taping measurements showed that horizontal movements were reasonably evenly distributed with no obvious concentrations of strain. The maximum tensile and compression horizontal strains recorded in any single span were 0.005% and 0.0085% respectively.

149

Monitoring of some significant pre-existing vertical cracks within the Palace commenced in January 1996 at the start of grouting operations. The changes in crack width throughout the remainder of the construction period and for six months thereafter were in the range of +0.5 mm opening and −1.5 mm closing. There was a clear annual cycle in the observations with the maximum widths being recorded in January/February and the minimum in August. Temperature induced variations were shown to be more significant than any change in crack width induced by the construction operations.

11   POST CONSTRUCTION BEHAVIOUR

Following the completion of construction in September 1997 the tilt of the Clock Tower continued to increase because of the ongoing consolidation of the London Clay. By the end of 1997 it was evident that, although the rate of tilting was clearly decreasing, the agreed Performance Control Level (PCL) of 1:2,000 would be exceeded within a matter of months. The compensation grouting facilities were still in place and could be used if necessary. However it was felt that a further episode of grouting might accelerate the movements again and be counterproductive. It was therefore decided to review the PCL in the light of the monitored behaviour to ascertain whether it could be relaxed. A thorough review of the historical monitoring data, background movements, performance during construction and available long-term settlement data from the JLE elsewhere in the Westminster area was undertaken (Harris, 2001b). A prediction of the probable increase in tilt was made in March 1998, 6 months after the end of construction and the final compensation grouting episode. This suggested that the tilt would increase to about 40 mm over a period of about 5 years.

Examination of the crack width measurements revealed the mechanism by which an increase in the tilt of the Clock Tower was reflected in increased crack widths within the structure. A detailed statistical analysis that took account of the effects of temperature change showed that for each 1 mm increase in tilt over the vertical gauge length of 55 m the average crack width at a high level in the building would increase by 0.07 mm. This correlation together with the predicted increase in tilt suggested an increase in crack width of just under 3 mm. An assessment of the form of the structure and of the existing cracking led to the conclusion that an opening of the existing cracks by up to 3 mm would not significantly affect the ease and cost of repair (i.e. the level of damage—see Burland, 1995). On this basis it was agreed that the PCL on tilt could be raised to about 40 mm.

Figure 13 shows the results obtained from the optical plumb measurements over the period of time since they were initiated in the early 1970s for the construction of the underground car park (Burland

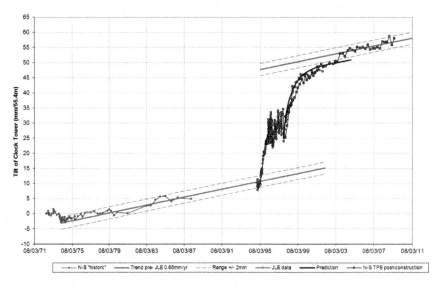

Figure 13.   The long-term northward movements of the Clock Tower prior to and subsequent to the construction on the Jubilee Line Extension underground railway.

and Hancock, 1977). As reported by these authors, the construction of the car park caused the tower to tilt about 2.5 mm to the south at a height of 55 m. The measurements show that in the 22 years following the completion of the car park the Clock Tower underwent a background rate of tilt to the north of about 0.65 mm per annum. The fluctuations around this trend of about ±2.5 mm are due to the seasonal and daily thermal effects mentioned previously.

Figure 13 also summarises the changes in north-south tilt of the Clock Tower during and subsequent to the construction of the new JLE station. It can be seen that, after the last grouting episode in September 1997, the trend has been a reasonably steady reduction in the rate of northerly tilt. Since about April 2003 the rate of tilt has reduced to the background rate of about 0.65 mm per annum established over a period of 22 years prior to the construction of the JLE. Taking account of this background rate it would appear that the magnitude of the tilt induced by the construction activities amounted to about 38 mm at the height of the clock face. It has been estimated that, had compensation grouting not been carried out, the construction-induced northward tilt would have been about 130 mm which would have caused significant damage to the Palace of Westminster.

12 DISCUSSION AND CONCLUSIONS

This paper describes the geotechnical measures that were used to protect the priceless historic buildings of the Palace of Westminster, including the Big Ben Clock Tower, from ground movements resulting from the construction of the nearby Jubilee Line Extension underground tunnels and the new Westminster Station.

The key conclusions that can be drawn from the work are as follows.

1. High-quality real-time measurements, particularly of the tilt of the Clock Tower, were vital to understanding its behaviour, to demonstrating the need for protective measures, and to controlling the compensation grouting that was implemented.
2. Before and during the works valuable insights into the effects of the construction works and possible responses of the buildings were gained from careful interpretation of appropriate numerical modelling.
3. The compensation grouting works were strictly regulated through a monitoring control office and proved extremely successful in controlling the settlement and tilt with considerable precision.
4. The understanding gained during the works led to an effective predictive model with which to estimate and monitor the time-dependent tilt of the Clock Tower.
5. Most of the observed movements within the Palace of Westminster concentrated on pre-existing cracks, which were very temperature dependent. Crack-width changes from seasonal temperature effects were comparable in magnitude to those associated with the ground movements due to tunneling and excavation.
6. More generally, while numerical modelling can be helpful in understanding mechanisms of behaviour, it is not possible to make precise predictions of the form and magnitude of ground movements due to excavation and tunnelling. These depend on a number of factors that are usually not known at the design stage, the most important of which are the detailed construction method and timing and their influence on the ground properties. Therefore it is essential to monitor the movements as they develop and thereby progressively refine understanding of both the ground movements and the response of the nearby structures impacted by them.
7. In this case the finite element analysis undertaken at the design stage was updated to take account of the revised construction methods and sequences. This revised analysis was calibrated against the observed settlements to give the best possible prediction of future movements.
8. If there is a risk of unacceptable damage due to ground movements, mitigation measures should be designed and put in place at the beginning of the work so that they can be implemented speedily and with a minimum of disruption.
9. For historic buildings the level of damage and the mitigation methods used to control it should be in accordance with the principles of conservation as far as is practicable.

151

REFERENCES

Attewell, P.B., Yeates, J. and Selby, A.R. (1986). *Soil movements induced by tunnelling and their effects on pipelines and structures.* Blackie, Glasgow.

Burland, J.B. (1995). Invited Special Lecture: Assessment of risk of damage to buildings due to tunnelling and excavation. *1st Int. Conf. on Earthquake Geotechnical Engineering*, Tokyo, 3, pp 1189–1201.

Burland, J.B. and Hancock, R.J.R. (1977). Underground car park at the House of Commons, London.: geotechnical aspects. *The Structural Engineer*, 55 (2), 87–100.

Burland, J.B., Standing, J.R., Linney, L.F., Mair, R.J. and Jardine, F.M. (1996). A collaborative research programme on subsidence damage to buildings: prediction, protection and repair. In: *Geotechnical Aspects of Underground Construction in Soft Ground*. R.J. Mair & R.N.Taylor (eds), Balkema, Rotterdam, pp 773–778.

Burland, J.B. and Viggiani, C. (1994). Osservazioni sul comportamento della Torre di Pisa. *Rivista Italiana di Geotecnica*, 28; 3; 179–200.

Bailey, R.P., Harris, D.I. and Jenkins, M.M. (1999). Design and Construction of Westminster station on the Jubilee Line Extension. *Proc. Instn. Civ. Engng, Jubilee Line Extension* 1999, 132, 36–46.

Carter, M.D., Bailey, R.P. and Dawson, M.P. (1996). Jubilee Line Extension, Westminster Station design. In:, *Geotechnical Aspects of Underground Construction in Soft Ground,* R.J.Mair & R.N.Taylor (eds), Balkema,Rotterdam, pp 81–86.

Clough, G.W. and O'Rourke, T.D. (1990). Construction induced movements of insitu walls. In: *Design and performance of earth retaining structures—Proc 1990 Speciality Conf.* Geotechnical Special Publication 25, ASCE, New York, pp 81–86.

Crawley, J.D. and Stones, C.S. (1996). Westminster Station—Deep foundations and top down construction in central London. In: *Geotechnical Aspects of Underground Construction in Soft Ground.* R.J. Mair & R.N.Taylor (eds), Balkema, Rotterdam, pp 93–98.

Harris, D.I. (2001a). Protective measures. Chapter 11 in: *Building response to tunnelling. Case studies from the Jubilee Line Extension, London*, Volume 1, *Projects and methods*. Burland, Standing and Jardine (eds.). CIRIA Special Publication 200. CIRIA and Thomas Telford, London, pp 135–176.

Harris, D.I. (2001b).The Big Ben Clock Tower and the Palace of Westminster. Chapter 18 in: *Building response to tunnelling. Case studies from the Jubilee Line Extension, London*, Volume 2, *Case Studies*. Burland, Standing and Jardine (eds.). CIRIA Special Publication 200. CIRIA and Thomas Telford, London, pp 453–508.

Harris, D.I., Mair, R.J., Burland, J.B. and Standing. J.R. (2000). Compensation grouting to control tilt of Big Ben Clock Tower. In: *Geotechnical Aspects of Underground Construction in Soft Ground.* Kusakabe, Fujita & Miyazaki (eds), Balkema, pp 225–232.

Harris, D.I., Pooley, A.J., Menkiti, C.O. and Stephenson, J.A. (1996). Construction of low–level tunnels below Waterloo Station with compensation grouting for the Jubilee Line Extension. In: *Geotechnical Aspects of Underground Construction in Soft Ground.* R.J. Mair & R.N.Taylor (eds), Balkema, Rotterdam, pp 361–366.

Rankin, W.J. (1988). Ground movements resulting from urban tunnelling; predictions and effects. *Engineering Geology of Underground Movement*, Geological Society, Engineering Geology Special Publication No. 5, pp 79–92.

Standing, J.R. and Burland, J.B. (2006). Unexpected tunnelling volume losses in the Westminster area, London. *Geotechnique*, 56, No. 1, pp11–26.

Stevens, A., Corbett, B.O. and Steele, A.J. (1977). Barbican Arts Centre: the design and construction of the substructure. *The Structural Engineer*, Vol 55, pp 473–485.

Geotechnics and Heritage – Bilotta, Flora, Lirer & Viggiani (eds)
© 2013 Taylor & Francis Group, London, ISBN 978-1-138-00054-4

Contributions of geotechnical engineering for the preservation of the Metropolitan Cathedral and the Sagrario Church in Mexico City

E. Ovando-Shelley
Instituto de Ingeniería, Universidad Nacional Autónoma de México, Mexico City, Mexico

E. Santoyo
TGC Geotecnia, Geotechnical Consultants, Mexico City, Mexico

ABSTRACT: In this paper we describe the project for the geometrical correction of Mexico City's Metropolitan Cathedral and Sagrario Church. In the first part of the project both temples were underexcavated between 1993 and 1998. The corrective deformations induced by this procedure nullified most of the differential settlement that accumulated during the previous 65 years, due to regional subsidence. Upon the end of underexcavation, the effects of regional subsidence reappeared and some of the corrections gained were lost. We also describe the second part of the in which the very soft clayey soils underlying the Cathedral and the adjacent Sagrario Church were hardened selectively by means of injected mortars, as a preventive measure to avoid the accumulation of future differential settlements. Observational data demonstrate that the procedure has been successful.

1 INTRODUCTION

Constructing Mexico City's Metropolitan Cathedral on extraordinarily soft soil was a formidable challenge back in 1573, when the building was started. Its creators took advantage of the experience gained by the Aztecs during construction of their Major Temple. In the case of the Aztecs, to the Mesoamerican tradition of superimposing new pyramids over the old ones during the festivities of the New Fire, they incorporated the practical need of adding successive construction stages to their buildings with the implicit purpose of concealing damage produce by differential settlements.

Master builder Claudio de Arciniega conceived an outstanding foundation for the Cathedral but even so, settlements brought about during the construction of the massive building compelled the succeeding architects to incorporate architectural ingenuity to mask misalignments. In 1630, Juan Gómez de Trasmonte erected the vaults and the transept. Luis Gómez de Trasmonte was appointed in 1656 to build the main dome. He was uncertain about the load bearing capacity of the transept columns and his suggestion of enlarging them was not followed. Starting in 1749, Lorenzo Rodríguez constructed the adjoining Sagrario (parish church) and he adopted a similar foundation system, but with a lesser quality. Damián Ortiz de Castro decided to repair the San Miguel chapel so it could bear the weight of the western bell tower and also began constructing the campaniles in 1780. Manuel Tolsá completed the Cathedral in 1813 after harmonizing the building and embellishing the dome. The construction process took 240 years.

The Cathedral and the Sagrario church have survived up to now thanks to restorations that have taken place over more than 300 years. Interventions have been increasingly complex due to the accumulation of structural damage and inclination, and the exposure to ever higher differential settlement rates. It is nearly ten years now since the end of the Project for the Geometrical Correction of the Cathedral and the Sagrario Church and for hardening its subsoil. It is very satisfactory to be able to state that, as verified with field measurements, the behavior of the religious complex improved very favorably after the successive application of underexcavation and selective soil hardening.

2 DESCRIPTION OF THE FOUNDATIONS

The Metropolitan Cathedral was built on part of the land covered originally by the Aztec Ceremonial Precinct (Ovando-Shelley and Manzanilla, 1997). Remains of structures corresponding to this pre-Hispanic site can still be seen under its foundation, Figure 1. As seen in Figure 2, the Cathedral has five naves: the central one bounded by 16 columns and divided by the choir; the two processional aisles running along the length of the church; and the two lateral ones occupied by chapels, that are in turn confined by the peripheral and perpendicular walls. The great central dome, 65 m high, is supported by four columns. The two towers are 60 m tall and the height of the temple is 25 m along the central nave, over an area 60.40 m wide and 126.67 m long with a total weight of 12,700 kN and an average contact pressure of about 166 kPa (see Figure 2).

The adjacent Sagrario is a church with a Greek cross layout whose walls at the four corners provide support to the vault; its dome rests on four columns. It covers a square area of 47.7 m by side, weighs about 3,000 kN and the average contact pressure is about 132 kPa, Figure 2.

Construction stages of the Cathedral. Construction of the Metropolitan Cathedral started in 1573 at the apse, under the direction of Master Builder Claudio de Arciniega, who had participated in the building of San Agustín Church and thus knew of the problems brought about by the underlying soft Mexico City Clays. The vaults were erected next and were completed around 1667 and the façade in 1675. Damián Ortiz de Castro finalized the towers in 1791 whereas Manuel Tolsá profiled the dome and joined the complex with a balustrade and pinnacles as a characteristic architectural feature. He completed the building in 1813.

The subsoil was initially reinforced by driving about 22,500 wooden stakes, 3 to 4 m in length. On top of them a masonry platform was built over an area of 140 by 70 m. This area is larger than the one

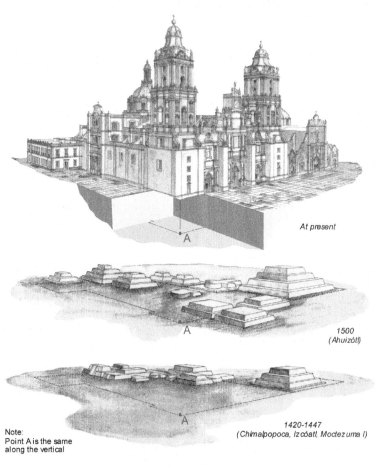

Note:
Point A is the same along the vertical

At present

1500
(Ahuizótl)

1420-1447
(Chimalpopoca, Izcóatl, Moctezuma I)

Figure 1. The Cathedral and the remains of the buried Aztec temples.

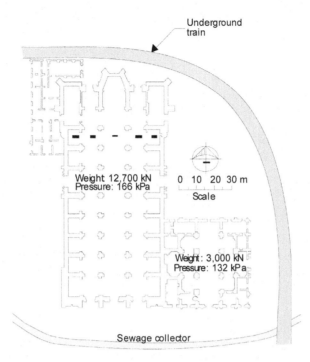

Figure 2. Dimensions and weights of the Cathedral and the Sagrario Church.

actually occupied by the Cathedral because it was originally conceived as a seven nave temple with four towers, one in each corner. The platform is 90 cm thick on average but it is thicker towards the south which suggests that the first builders added thickness at that particular zone to compensate differential settlements that became apparent since the earliest stages of its construction. A grid of inverted beams was built on top of the platform with masonry as well, 3.5 m in height, 2.5 m wide and as much as 127 m long, to receive the walls, pilasters and columns, as illustrated in Figure 3. The top part of the platform had the same level as the Plaza Mayor (Main Square) and the grid of beams was 3.5 m above this elevation which clearly indicates that Master Builder Arciniega expected large-magnitude settlements to occur.

Other religious buildings were built around the Cathedral. The most remarkable structure is the parish church known as the Sagrario, built on top of the pyramid of the Aztec sun god, Tonatiuh (also illustrated in Figure 3). For the construction of the Sagrario, Lorenzo Rodríguez used the same foundation system as in the Cathedral, reinforcing the soil with short woodpiles having a smaller diameter. On top of them a masonry platform was built but with lesser quality materials. The Sagrario was partially founded on the Cathedral's foundation platform and its western wall is common to both structures. The construction of the Sagrario stretched from 1749 to 1768. The Bishopric was built later, as well as All Souls Chapel (Capilla de las Ánimas) and the Seminary which was demolished in 1938.

Cathedral surroundings. In 1968 the semi-deep sewage collector "5 de Mayo" was built at a depth of 16 m along the southern facade of the Cathedral and of the adjoining Sagrario. It has been inferred from Piezometric measurements that the collector is permeable and that water seeps into it from the subsoil, particularly at the southeastern zone of the Sagrario. Construction of Line 2 of the subway system (Metro) also started in 1968 and its cut-and-cover tunnel also acts as a drain at the north and east sides of both churches (Figure 2).

Settlements during construction. Consolidation of the subsoil induced by Aztec temples and structures pre-existing at the site produced differentials in compressibility of the subsoil clay strata which in turn, caused differential settlements since the beginning of the construction. These deformations brought about structural misalignment that was compensated as construction progressed

155

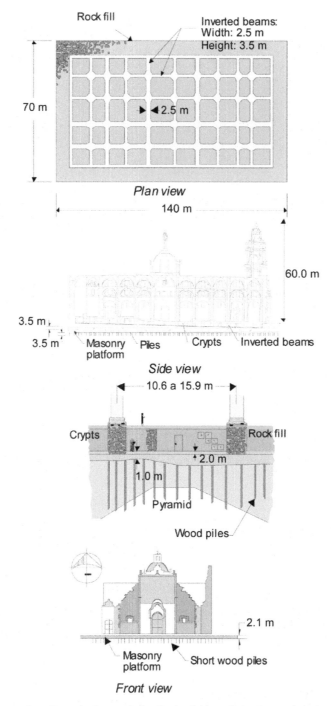

Figure 3.   Foundations of the Cathedral and the Sagrario Church.

by modifying the heights of columns and walls in order to level the springing of the vaults. Architectural contrivances as the introduction of variable heights in the cornices and wedged quarried blocks at the two towers were used to disguise some of the visual effects of settlements. After analyzing the geometrical details of the monument it was demonstrated that during construction of the Cathedral, and prior to the completion of the vaults, column C-9 accumulated a maximum differential settlement of 85 cm with respect to the plinth of pilaster C-3 in the polygon that forms the apse (see Figure 4).

156

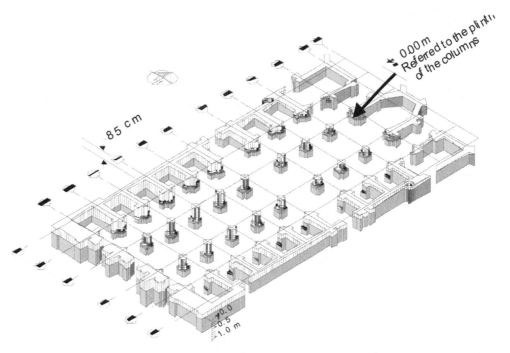

Figure 4. Enlargement of column shafts due to differential settlements during construction.

3 INTERVENTIONS IN THE FOUNDATIONS

In 1929, the Technical and Conservation Commission for the Cathedral appointed architects Manuel Ortiz Monasterio and Manuel Cortina García to make a structural evaluation of the Cathedral because settlements had caused alarming damages. As a first measure, it was decided to demolish the seminary, to unload the east zone.

First intervention in the Cathedral. Architects Ortiz Monasterio and Cortina García decided to empty the earth fills from the cells of the inverted beam grid supporting the Cathedral with which the average contact pressure decreased from 143 kPa to 108 kPa, i.e. a reduction of 25%. The project also considered the reinforcement of the masonry inverted beam grid with reinforced concrete (Figure 5). A few years later crypts were installed in the empty cells and gaps were opened through the beams to form access aisles. The masonry elements were reinforced with structural steel beams that were supported by a concrete slab with an approximated thickness of 50 cm (Figure 5) that was only built at the east and west sides of the transept. Finally, the wooden floor at the parish level was replaced by a reinforced concrete slab with a construction joint left along the western side column axis.

First intervention at the Sagrario church. An attempt to underpin the Sagrario Church took place in the forties, with 25-cm diameter woodpiles. In addition, the parish floor was reinforced with a concrete slab supported by a grid of steel beams. Subsequently, between 1960 and 1964 another underpinning system was tried at the using concrete piles driven in one meter lengths. Many of the top parts of such piles can be observed at the cells under the Sagrario; it is evident that a large amount of them could not be driven.

Second intervention in the Cathedral and the Sagrario. In 1972, the Secretaría del Patrimonio Nacional (SPN) commissioned Mr. Manuel González Flores to study settlements in the Cathedral. He recommended the installation of control piles to "reduce load demands on the foundation in about 25%, to adjust the descent of the building with respect its surroundings and to achieve a uniform distribution of settlements". His proposal was to install 280 piles in the Cathedral, mainly in its southern part (Figure 6) and he did not specify the exact number of piles needed in the Sagrario. Practical difficulties forced him to install them where possible and, hence to increase its number to 390 at the Cathedral; 129 piles were installed at the Sagrario.

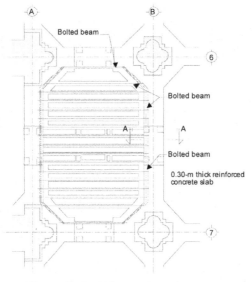

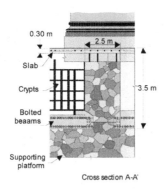

Cross section A-A'

Figure 5. First intervention in the foundations in 1940 due M. Ortiz-Monasterio.

Classification of the foundation piles. The piles were classified as reliable and as inefficient on the basis of data reported in the project logbook. The former have their tips properly supported by the First Hard Layer. The latter may be separated into four groups: a) short piles whose tips do not reach the First Hard Layer and work as friction piles; b) long, inclined or broken piles were those whose reported lengths were larger than the depth required to reach the First Hard Layer; c) those installed where they were not actually required, unnecessary piles; d) long friction piles installed in the outer edge that, given the technique used to drive them, may have punctured the First Hard Layer from the atria. A statistical analysis made in 1989 concluded that only 27% of the piles are properly supported by the First Hard Layer, and at the Sagrario only 11% of the piles fulfill such condition (refer to Figure 6).

Comments. The first and second interventions were conceived without having a proper knowledge of the characteristics of the subsoil and its behavior. In the first intervention the benefits of removing soil to uniform settlements turned out to be quite limited because regional settlement soon compensated it and, the weight of the soil removed was nearly the same as the weight of the crypts; furthermore, in building the crypts, the inverted beams were weakened.

Regarding the control piles, assuming that only 103 of those installed at the Cathedral and 14 of those placed in the Sagrario are reliable and considering that their individual bearing capacity was 100 t, it follows that a total bearing capacity of only 11,700 t is available, a third of the capacity originally expected and only 7.5% of the total weight of the complex, 157,000 t, which is obviously insufficient to modify the behavior of the foundation of each of the churches.

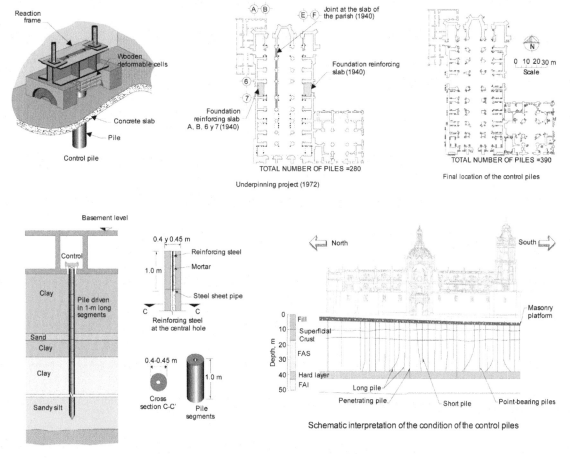

Figure 6. Second intervention in the foundation by M. González-Flores in 1972 with control piles.

Control piles have proven to be useful for underpinning rigid modern buildings that can allow continued maintenance operation for these piles which imply the temporary removal of the reaction frames and eventually, the trimming of the pile caps. The large dimensions and structural flexibility of the Cathedral and the Sagrario deem the capabilities of control piles for controlling settlements in the long term as imperceptible.

4 SUBSOIL CHARACTERISTICS

Geological, geotechnical and historical information ratified that the Cathedral was erected on a natural islet which was only a small promontory with a spring known to the Aztecs as Toxpálatl, that existed under what is presently the west atrium. Geotechnical explorations were carried out to define in detail the underlying stratigraphy at the sites as well as subsoil properties with emphasis on the compressibility of the materials. Preliminary studies performed in 1989 included 21 cone penetration tests (CPT tests) as well as two borings with continuous undisturbed sampling. In the course of the construction of 32 shafts in 1993, 29 additional CPT tests were made.

Stratigraphical profile. The stratigraphical profile shown in Figure 7 was produced from the results of three CPT borings performed in front of the Cathedral and of the Sagrario. As seen there, the soil at the boundary between both churches is stronger because it corresponds to the zone that has received the heaviest load transmitted by the Aztec temples, by an archaeological fill, and by the two heavy Colonial structures. Towards both ends of the profile penetration resistance reduces almost by a half and compressibility increases. This condition induced the tilting of the southern part of the Cathedral towards the west whereas the Sagrario is inclined to the east. This deformation pattern explains

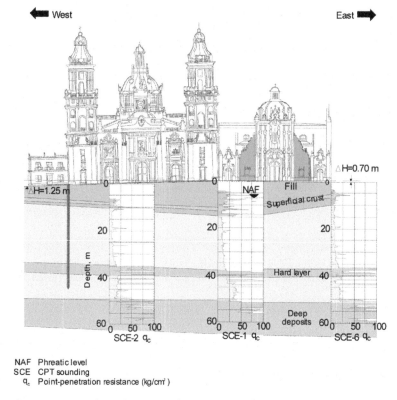

Figure 7. Stratigraphy at the site and surficial differential settlements.

the apparition of fissures at the west side of the Cathedral and at the east of the Sagrario. The same figure also shows the thickness and depth of the most relevant strata found in the soil sequence at the site.

Subsoil deformations. The depth of the contact between the natural shallow crust and the soft clays was defined from information derived from the CPT tests. That surface was originally flat but as a result of the consolidation induced by the Aztec pyramids it underwent depressions as deep as 10 m, as sown Figure 7. This is why the site was leveled with artificial fills to shape a new initial plane before the construction of the Colonial churches.

One-dimensional compressibility tests demonstrated that loads applied by the former pre-Hispanic constructions were removed at some parts, although in some other areas they were subsequently increased by the weight of the Cathedral and of the Sagrario. This complex load history brought about the heterogeneity in the conditions and properties of the subsoil that was detected in the field and laboratory tests, as illustrated schematically in Figure 8.

Pore pressure measurements in 1990. Pore-water pressures at seven depths were measured in a piezometric station installed at the southern atrium of the Cathedral down to a depth of 63 m. It can be observed in Figure 9 that between 0 and 20 m in depth, pore pressure is nearly hydrostatic; below this depth a pressure loss of about 180 kPa was noted at the First Hard Layer, 38 m deep, and of 200 kPa at the Deep Deposits, 53 m deep.

Estimates of future water pressure trends. Considering that water extraction from the subsoil will continue indefinitely, pore pressure distribution recorded at piezometric station EP-1 can be expected to slowly decrease in the future and that pore water may eventually define a hung aquifer formed by the infiltration of rainwater and by seepage from potable water and sewage mains. With these hypotheses two predictions of the piezometric variation were established, (Figure 9):

Prediction 1. It is feasible to imagine a suspended body of "trapped water" located between 6 and 25 m in depth as well as a hydrostatic distribution underlying the former. This assumption implies a decrease of the hydraulic pressure down to a value of 180 kPa at the Upper Clay Formation.

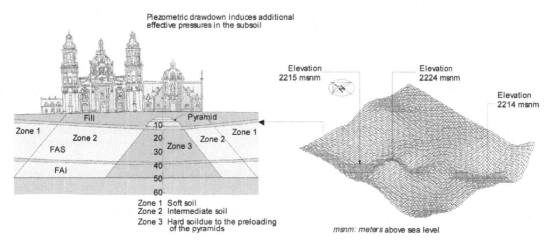

Piezometric drawdown induces additional
effective pressures in the subsoil

Elevation 2215 msnm
Elevation 2224 msnm
Elevation 2214 msnm

Fill
Pyramid
Zone 1
Zone 2
Zone 3
Zone 2
Zone 1
FAS
FAI

Zone 1 Soft soil
Zone 2 Intermediate soil
Zone 3 Hard soil due to the preloading
        of the pyramids

msnm: meters above sea level

Top surface of the upper clays as deformed by the
weight of the old pyramids and of the Cathedral

Figure 8.   Effect of the buried pre-hispanic structures on soil properties and deformations.

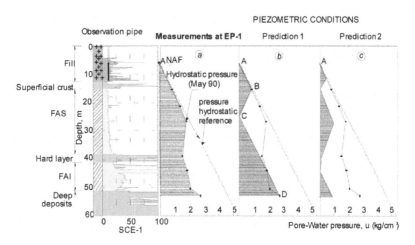

Figure 9.   Predictions of piezometric drawdown.

Prediction 2. It can also be assumed that two hung water levels will be formed, one of them between 6 and 13 m in depth and the other from 16 to 38 m. This implies pore pressure drops at such depths of 80 and 180 kPa, respectively. Furthermore, beyond a depth of 45 m a hydrostatic distribution may also be reached.

5   REGIONAL SUBSIDENCE

Regional subsidence, which induces damages in Mexico City, is the consequence of groundwater extraction from underlying aquifers and from pore water pressure reductions there (Carrillo, 1948). The aquifer has two major characteristics: a) it is constituted by pervious materials such as sand, sandy silt or gravel; and b) it is confined by low-permeability clays. As water pressure in the aquifer decreases, a gradual reduction of the pressure in the water filling the pores of the clays also occurs. Depending on the thickness and the permeability of the clay, a change in water pressure in the aquifer produces deferred changes in the pore-water pressure of the low-permeability materials that may last even decades before reaching a new state of equilibrium. Water will also flow downwards very slowly from the clay into the aquifer.

161

When the clays are saturated, as it is very closely the case in Mexico City, the volume of water expelled is proportional to sinking observed at the ground surface. Pressure changes undergone by the pore-water pressure in the clay increase the stresses acting effectively in the solid phase of the soil.

Recorded settlements. In 1860 Javier Cavallari performed the first leveling between the Cathedral and what he thought was a fixed basalt outcrop at the atrium of a church at a nearby town, Atzacoalco. Cavallari also made the second leveling about ten years later and the third one was carried out by Roberto Gayol in 1892, from the same rock outcrop to the lower tangent of the Aztec Calendar (TICA) that used to be attached to the Cathedral's west tower. It was later established that the outcrop was in fact a

Atzacoalco bench mark: basic reference for all topographic levelings in Mexico City.

One of the reference mosaics placed in 1906 downtown Mexico City.

Figure 10.   Topographic references in Mexico City.

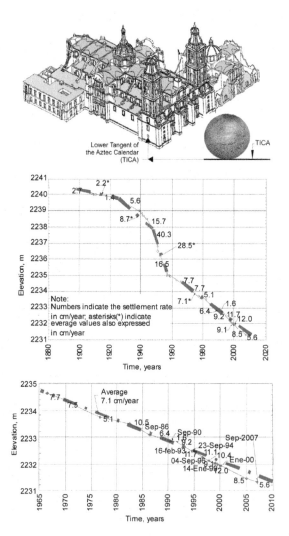

Figure 11.   Total settlements measured at the lower tangent of the Aztec Calendar.

162

loose block and, hence, not a reliable reference. Another benchmark was installed in 1937and a newer one exists since 1959, Benchmark No. 251, which became there after the basic reference for topographic jobs in Mexico City, Figure 10.

The effects of regional subsidence can be illustrated by analyzing the development of settlements at the TICA historical reference. As shown in Figure 11, this reference settled more than 8 m during the 20th century, and approximately 2.6 m over the last 35 years. The graph highlights the most significant values of yearly settlement rates. Between 1965 and 1990 subsidence varied almost linearly, approximately at a rate of 7.1 cm/year. In the leveling made in 1991, this rate had decreased to only 1.6 cm/year and subsequently, because of the effects of underexcavation at the Cathedral, it increased to 10 cm/year from 1991 to 2000. Rates of 7.5 and 6.1 cm/year have been measured during 2006 and 2007, respectively.

Settlement distribution within the soil. Deep bench marks, 40, 60, 80 and 100 m deep, were installed in the Cathedral's atrium with a twofold purpose: a) to measure total settlements, and b) to determine the distribution of settlements within the subsoil. These benchmarks are constructed with twin concentric pipes. The internal one act as a reference mark and therefore it is continuous and rests at the selected depth. The external pipe is compressible and, hence, it absorbs axially the vertical deformations undergone by the soil between the surface and the depth of the benchmark. The inner tube remains free, i.e. it is not affected by buckling as are conventional bench marks built with rigid outer pipes. Friction forces acting against the inner pipe are in fact eliminated.

Settlements measured at the deep benchmarks are shown Figure 12 as well as the contribution in percentage of the major compressible strata to total settlements. In 1991 before geotechnical work in the Cathedral began, the Upper Clay Formation contributed with 54%, the Lower Clay Formation and the deep silty clays of the former third lake, with 46%; settlements below 80 m were nil. At the time

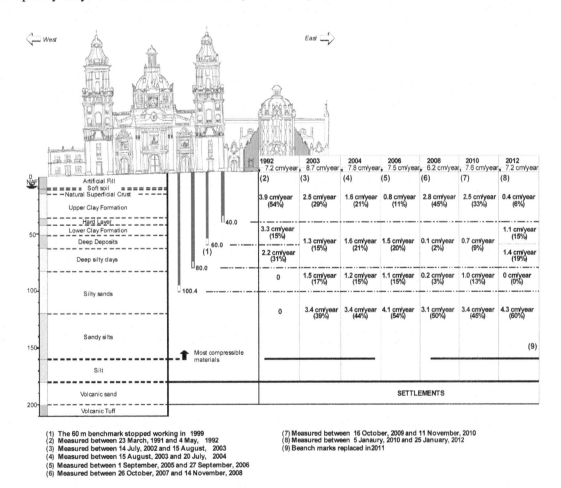

(1) The 60 m benchmark stopped working in 1999
(2) Measured between 23 March, 1991 and 4 May, 1992
(3) Measured between 14 July, 2002 and 15 August, 2003
(4) Measured between 15 August, 2003 and 20 July, 2004
(5) Measured between 1 September, 2005 and 27 September, 2006
(6) Measured between 26 October, 2007 and 14 November, 2008

(7) Measured between 16 October, 2009 and 11 November, 2010
(8) Measured between 5 Janaury, 2010 and 25 January, 2012
(9) Beanch marks replaced in 2011

Figure 12.   Distribution of vertical deformation within the clayey soils under the Cathedral.

those data were disquieting because they proved wrong the ancient hypothesis that considered that the compression of the uppermost clays was the sole contributor to regional subsidence. The measurements after that year show that the contribution of the upper clay formation has decreased in general, especially from 2008 onwards. Regarding the lower strata, the measurements are even more alarming because they show that the contribution of the so called deep deposits has increased steadily and it now accounts to 60% of the observed settlement, i.e. the regional consolidation is now affecting the soils that underlie the upper, more compressible clays. It is urgent that these observations be verified with measurements at other sites in the city.

6 GEOTECHNICAL DIAGNOSIS

Accumulated settlements in the Cathedral over 419 years, from the beginning of construction until the end of 1989, generated a differential settlement of more than 2.4 m between the apse and the western tower (Figure 13). Development of deformations there are the sum of two components: a) consolidation induced by the weight of the pre-existing Aztec temples and of the subsequent Colonial structure; and b) regional subsidence of the city. The latter has been the most important factor for the development of differential settlements during the last 150 years; between 1907 and 1989 it induced a differential settlement of 87 cm in the west tower with respect to a brass bolt installed by Roberto Gayol in 1907 at the plinth of a pilaster at the apse, which has ever since been considered as the zero reference, as seen in the upper part of Figure 13.

In order to assess the effect of the regional subsidence in the development of differential settlements at the structures, several precision topographic surveys were carried out at the Cathedral and the Sagrario during the stage of preliminary studies. Topographic levelings were performed at the plane of the plinth of the columns supporting the Cathedral as a continuation of measurements of this surface that have been carried out since 1907.

Figure 14 shows recorded annual settlement rates obtained from measurements made in the period comprised between January 7 and September 2, 1991before underexcavation. From the figure it is possible to infer the geometric deformations suffered by both churches during that time and it represents the trends that would have been observed, had underexcavation not been carried out. As an example, the western tower settled at a rate of 12 mm/year with respect to the central part of the nave; the southeastern corner of the Sagrario was settling 16 mm/year with respect to its central part, and

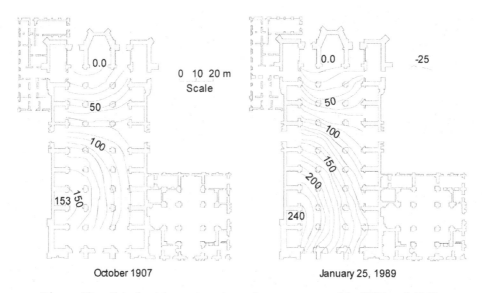

October 1907   January 25, 1989

Figure 13.   Equal settlement contours in cm, measured in 1907 and 1989.

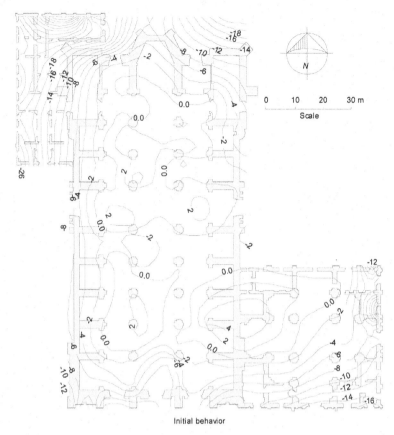

Initial behavior

Figure 14. Equal settlement rate contours (mm/year) measured between 7 January and 2 September, 1991.

the vertical deformation of the museum building was of 26 mm/year with respect to the zero reference point.

Estimation of initial settlements. Settlements induced by the Aztec pyramids at the zone where the Cathedral and the Sagrario were later built were estimated assuming the probable thickness that the soil strata had under both churches prior to the construction of the pre-Hispanic structures, following the same method used by Mazari et al., (1985) to reconstruct the stress and strain history of the subsoil under the Great Aztec Temple. It was estimated that the weights of the Aztec pyramids induced compressions in the soft clays that ranged between 7 and 13 m. These values agree reasonably well with those deduced from the stratigraphical interpretation of CPT test results.

Prediction of future differential settlements. Long-term settlements forecasted using traditional soil mechanics procedures and methods were carried out assuming that the churches would be left as they were in 1989, providing a panorama of the consequences that would have been faced, had they not been treated. As mentioned before, future settlements at the Cathedral and the Sagrario depend on the evolution of the pore-water pressures in the clay deposits and on the initial pore pressure distribution.

Two hypotheses were assumed for the future hydraulic conditions likely to prevail in the subsoil, Figure 15. Prediction 2 leads to more pessimistic estimates of future differential settlements (for example, 3.2 m at the base of the western bell tower). In the case of the Sagrario, the average differential settlement between the central zone and the corners could be of 1.2 m, retaining the same zero reference.

From the results presented above, it was concluded that a large magnitude earthquake such as the one that occurred in 1985 could induce a stress condition that could be seriously risky to the stability of the churches, particularly that of the western tower.

165

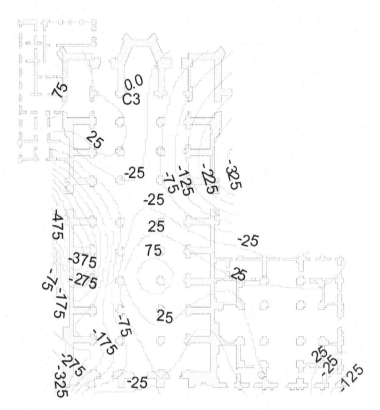

Figure 15. Contours of equal differential settlements estimated assuming the pore pressure distribution of hypothesis 2.

7 SOLUTIONS STUDIED

The following five possible solutions for correcting historic differential settlements and to reduce future differentials were studied:

Piles supported on the First Hard Layer. Their aim was to produce uniform settlements of the foundation and the soil mass under the Cathedral by driving 1500 point-bearing piles to the First Hard Layer, capable of supporting through negative skin friction (down drag) the surrounding ground and the Cathedral itself. This solution would increase the stiffness of the soil and the structure enabling them to tolerate the sinking of the Upper Clay Layer. It would also induce the apparent emergence of the structure with respect to the surrounding ground level, as it occurs in other conspicuous buildings in the city.

Piers supported by the Deep Deposits. With this solution, the settlement of the structure would not depend on the sinking of the two clay formations. Pier tips would be supported by the Deep Deposits and would be connected to the foundation by means of mechanical devices to correct existing tilts and to avoid the accumulation of further tilting in the future. Conceptually, these elements would be similar to the control piles described before; 240 piers are needed to carry all the weight of the Cathedral and down drag generated when soil settles due to the regional subsidence.

Underexcavation in soft clays. This technique was suggested as a means for correcting some of the inclination of the Tower of Pisa by the Italian engineer, Ferando de Terracina, as illustrated in Figure 16 (Terracina, 1962). For the Mexican Cathedral it required drilling 10 cm diameter "micro-tunnels" that would close due to plastic deformation or failure of the soft clays; successive opening and closure of the tunnels would gradually induce corrective settlements until reaching the deformation targets fixed according to structural considerations. This is why the project was named the Geometrical Correction for the Cathedral and the Sagrario Church. Underexcavation was later applied at the Tower Pisa later, as discussed in several other papers (e.g., Jamiolkowsky et al, 1999).

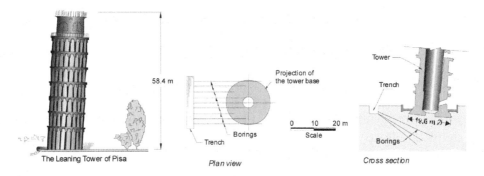

Figure 16.   Terracina's (1962) proposal for underexcavating the tower of Pisa.

Pore-water recharge. Since the origin of regional subsidence is hydrological, the artificial recharge of water into permeable subsoil strata was studied. The brief experience gained with this technique at the National Palace was analyzed. Implementing it at the temples, would require 46 injection well and an impervious cutoff wall along the perimeter of both temples. Estimates showed that this would control 69% of the settlements provided water injections remained permanently; otherwise settlements would inevitably accumulate again.

Underpinning with micropiles. "Pali radice" or inclined and vertical small-diameter micro-piles were also studied. Intertwined inside the clays, these elements create hard blocks that transfer loads to the deeper strata. This solution would require an enormous amount of such elements and therefore, impossible to install.

Comparing the solutions. The five options were analyzed and compared critically. The solutions were evaluated making reference to the structural goals, interference for the usage of the temples, time of execution, budget, and probable contingencies.

It was concluded that intervening the temples applying the Underexcavation Method was the solution that best fulfilled the expectations for achieving the Geometrical Correction for the Cathedral and the Sagrario Church. Some of the specialists had doubts about the applicability of that technique to masonry structures, since the examples they examined were only related to reinforced concrete structures.

Experimental underexcavation. Underexcavation trials were performed in a smaller masonry structure whose architecture is similar to the Cathedral's and was also built shortly after the Spanish Conquest (the temple of San Antonio Abad). Trials were aimed at overcoming uncertainties associated then to the technique and to prove its feasibility. Underexcavation was performed from the bottom of three access shafts, 10 m deep; it induced rigid body movements and torsions and it was also demonstrated that the process could be precisely controlled. Members of the International Consultants Committee examined the trial and accepted its validity to go on with underexcavation at the Cathedral and the Sagrario Church.

8   UNDEREXCAVATION AT THE CATHEDRAL AND THE SAGRARIO

Underexcavation was devised to reduce differential elevations and tilting induced by differential settlements. It involved lowering the high parts with respect to the low points through the slow and controlled extraction of soil from the bearing strata. Three specific tasks are necessary to apply the method: a) the construction of access shafts; b) the punctual drawdown of the phreatic level; and c) underexcavation or controlled extraction of small portions of soil until removing a pre-established volume. The two first operations are preliminary; the third one constitutes the corrective geotechnical procedure itself.

Preparatory work. It began by excavating 32 access shafts lined with reinforced concrete. Their number and distribution were determined applying analytical and numerical methods. Their bottom varied between 14 and 25 m, was taken down to top of the Upper Clay Formation. Four small diameter point wells were installed inside each of the shafts to gradually drawdown the phreatic level and to prevent bottom failure; it operated locally throughout the whole soil extraction process to keep the water level below the bottom of the shafts and, hence, to allow underexcavation of the clay.

Underexcavation details. Soil was extracted from the soft clay located at the boundary of the Upper Clay Formation, the morphology of which was illustrated in Figure 7. At the most, 50 radial borings penetrated into the soil In each shaft, in lengths ranging 6 to 22 m. Figure 17 shows the layout of the shafts, an illustrative cross section of one of them and a profile of the lengths penetrated with underexcavation, as well as a sketch of the closure of the borings to induce the required settlements. Soil was extracted driving steel thin walled samplers with a hydro-pneumatic cylinder. Boreholes 10 cm in diameter were inclined 20° and a remolding tool was sometimes used to accelerate their closure.

Correction targets. Figure 18 shows the targets to be achieved by applying the underexcavation method. One of them was proposed by Dr. Fernando López Carmona, and the other one by Dr. Roberto Meli Piralla who performed, respectively, graphic-analytic and numerical structural analyses. The correction targets derived from these analyses were:

a) To close and rotate the lateral walls in order to strengthen the "confining belt" formed by the walls along the perimeter of the temple and along the sides of the chapels.
b) To lower the Cathedral's apse 80 to 95 cm, in a rigid body movement.
c) To lower the Sagrario's north side 30 cm in a rigid body movement.

A geometrical control model was devised to program these fundamental goals. The model was compared step by step with topographical measurements at the parishioners' level and with convergence measurements and conventional and electronic plumbs.

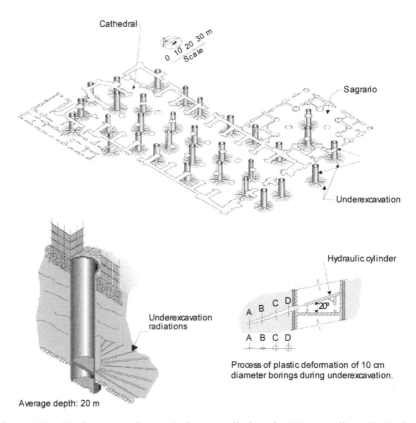

Figure 17.   Underexcavation technique applied at the Metropolitan Cathedral.

Structural bracing. Underexcavation was executed with the assistance of a preventive bracing system capable of controlling unexpected deformations and of preventing structural damages, Fig. 8. This system operated during the whole process and was adjusted periodically to the gradual changes induced by underexcavation. It never supported the totality of its design load.

Underexcavation control. The weight and the moisture content of the material extracted were accurately and rigorously monitored during underexcavation. Undisturbed soil samples were also retrieved for laboratory testing to obtain their mechanical properties. Soil extraction began in August, 1993, and finished in June, 1998; 4,220 m³ were removed in about 1,451,000 extraction operations. Underexcavation stopped once the structural targets of the project were achieved and, thus, the religious complex was once again exposed to the action of differential settlements of the subsoil (see Figure 19).

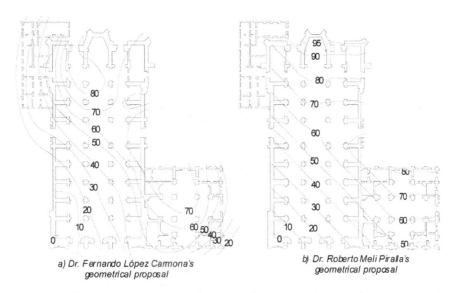

a) Dr. Fernando López Carmona's geometrical proposal

b) Dr. Roberto Meli Piralla's geometrical proposal

Figure 18.   Contours of proposed target settlements in cm.

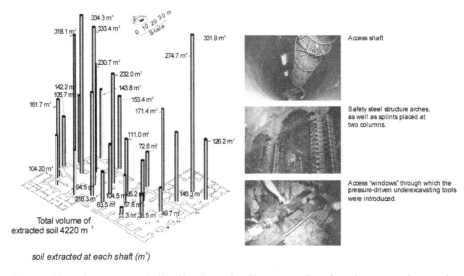

Figure 19.   Amount and distribution of soil extracted and underexcavation tools.

9 GEOMETRICAL CORRECTION ACHIEVED

Underexcavation eliminated differential settlements accumulated over the previous 65 years as a result of regional subsidence by the end of June 1998. The maximum correction induced was 92 cm, between the apse and the southwestern corner. However, as seen in Figure 20, towards September 1999 the maximum corrective settlement reduced to 88 cm and to 30 cm at the Sagrario. This difference from 92 to 88 cm is due, as discussed previously, to the fact that upon the end of underexcavation and the stoppage of the pumping operations, the effects of regional subsidence returned and, as a result, part of the corrective settlements that had been achieved was lost. Historical differential settlement between points C-3 and B-11 changed from 243 cm in 1989 to 174 cm in June 1998. The angular correction between these two points was 26.3′.

Geometry of corrective settlements. The configuration of corrective settlements is represented with meshes as those depicted in Figure 21. The upper mesh shows cumulative and corrective differential settlements as of September 1999. The lower part of the figure presents the same corrective settlements but referred to a horizontal plane, i.e. without considering historical differential settlement. The shape and distribution of the corrective settlements that reached a maximum of 88 cm, is also shown.

Corrections in the towers. Topographic surveys evinced differential elevations that were originally present in the main façade and the towers. Both towers were monitored periodically starting in October 1993 using electronic plumbs with a sensitivity of 0.1 mm and 35 m long steel wire lines (precision = 1/35,000 or 5.9″). Regarding the influence of underexcavation and mortar grouting, the correction achieved by April 2000 at the west and east towers consisted in rotations and displacements of 28.7 and 27.9 cm, respectively, along a northeast direction, equivalent to correcting 26.3 and 24.6% of pre-existing tilts in the west and east towers, which were initially inclined 109 and 113 cm, respectively.

Reference plumb line. Seeing with the naked eye the corrections that were in course was not possible and that prompted the installation of a large plumb line at the Cathedral's central dome, to make its movements apparent to visitors. Figure 22 presents the reconstruction of the path followed by the plumb line in the course of time. It is interesting to point out the change in the direction of its trajectory induced by differential settlements that resulted from the extraction of water from underground

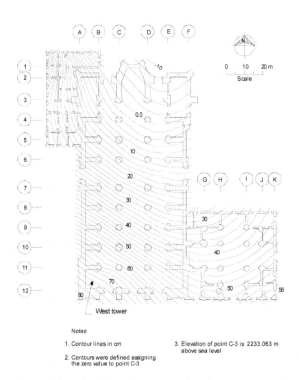

Figure 20. Differential correction achieved with underexcavation from 25 October, 1991 to 20 September, 1999.

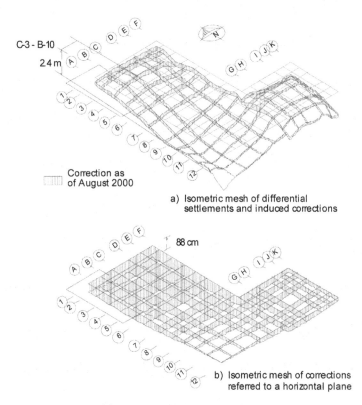

Correction as
of August 2000

a) Isometric mesh of differential
   settlements and induced corrections

b) Isometric mesh of corrections
   referred to a horizontal plane

Figure 21. Geometry of corrective settlements is induced with underexcavation.

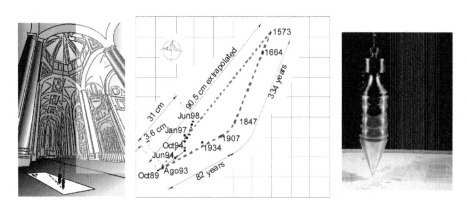

Figure 22. Reconstruction of the path followed by the dome.

aquifers, since the end of the 19th century. From October 1989 to October 1994 corrective movements in the dome made it lean towards the northeast; subsequently, after adjusting the underexcavation program and up to June 1998, it rotated mostly towards the north. The total correction of the dome's tilt was 31 cm which is equivalent to a corrective angular rotation of 25.3′.

Structural damage. Measurements showed that the vault rose a few centimeters during the early stage of underexcavation, a fact that indicated that the confinement provided to the vault by the walls was restored. The underexcavation sequence was adjusted later to produce a second type of movement, rigid body displacements towards the northeast. Corrective settlements contributed to the closing of the cracks and to reducing tilts in columns. However, new cracks developed and others that already existed widened. Also, an ashlars stone fell off a window and plastering dropped from some points as well. Nonetheless, damages were smaller than those expected at the beginning of the project.

Structural monitoring. The condition of the structure was recorded and logged by means of plumb lines and conventional deformation gages; later, a continuous monitoring electronic system was

171

installed, including accelerometers inside the temple and in the free field. Numerous analyses showed that the safety conditions of the churches were at no time at all in the course of the project in a situation of risk. The shoring, the confinement reinforcing at the columns (splints), and the turnbuckles that were installed at the roof acted as a protection to the structures against possible damage, as well as a steel net along the central nave to protect the parishioners. The condition of the columns was the most critical aspect regarding the safety of both churches and it was therefore decided to grout them to improve their safety factors. Since the harmful effects of regional subsidence reappeared when underexcavation stopped in June 1998, a small fraction of the corrections achieved was gradually lost but, in the middle of 2000 the beneficial effects of mortar injections in the subsoil began to be evident.

10   VALIDATION OF MORTAR INJECTIONS

Differential settlements at the former Teatro Nacional, presently the Palace of Fine Arts (Figure 23), were first noticed in 1906 when its foundation slab was being constructed. This prompted the injection of grouts into the soft underlying clays from 1910 to 1925. Initially cement grouts were used and later, fluid mortar, possibly made with lime and sand,. Those injections were intended to arrest settlements, although to no avail. It should now be acknowledged that grouting was great success because, even though settlements were not stopped, they became uniform (see Santoyo et al, 1998).

Many engineers questioned whether grouting the subsoil of the Teatro Nacional was indeed effective. Soil grouting was also misinterpreted because it was thought that the soil was being impregnated with the cement grout. During that time, regional subsidence in the city had not been acknowledged. The case lost momentum, political unrest of those years diminished interest on the topic, and finally the theater was left unfinished until 1934 when its construction concluded. The technical information was filed and the unfair remark that "grouting failed to serve its purpose" was the only judgment that remained about this case. This almost forgotten experience is a remarkable precedent of the method developed in the 1990s to modify the compressibility of the subsoil under the Metropolitan Cathedral.

Theoretical and experimental studies. Theoretical and experimental research into the effect of mortar grouts injected into soft clays to reduce selectively their compressibility began in 1997. The technique was evaluated from the results of field trials carried out at the former Texcoco Lake bed, and with complementary laboratory tests and numerical simulations. Field trials took place to identify the parameters governing hydraulic fracturing of Mexico City Clays; different combinations of pressure, volume flow, viscosity and shear strength in the mortars were tested. Hydraulic fracturing has been investigated in extensive theoretical and practical research in many countries, but it had never been studied formally in these clays, albeit it had been applied informally, with no technical support Experience with this type of injections in Mexico City was summarized by Ovando and Santoyo (2003).

Vertical mortar sheets. A 4.2 m diameter shaft was excavated to a depth of the treatment layer, to verify the formation, length and thickness of the cement/lime mortar sheets, Figure 24. Its walls were stabilized with reinforced shotcrete. Point wells were also installed to allow the excavation of the shaft to a depth of 6.15 m. By advancing the excavation in 50 cm stages it was possible to observe in full detail the grout sheets that covered lengths from 1 to 3 m, as illustrated schematically in Figure 24. These tests showed that fluid mortar injected into the clay produces fissures and cracks along planes whose orientation depends on the initial in situ stress state, as expected. Mortar injected into the clay penetrates in the fissures thus forming vertical sheets.

Rigid inclusions. Trials on the formation of mortar nuclei, more recently called rigid inclusions, were also performed. The inclusions were made opening a 23 cm borehole stabilized with slurry generated during the perforation. A permeable polyester fabric is then introduced into which fluid concrete was then cast. The integrity of the inclusions was verified with exploratory excavations performed afterwards and in which it was also revealed that the mean diameter in the inclusions was 29 cm, which implies a 26% radial expansion in the borehole diameter.

Grout tests at the Cathedral. A grouting experiment commenced in October 1997 at the Cathedral's west atrium to assess the feasibility of this technique to arrest future differential settlements. The presence of grouted mortar in the clay mass was verified with undisturbed samples retrieved with a 40 cm diameter rotation sampler driven to a depth of 12.5 m, Figure 25.

Photograph taken in December 1906, when the differential settlement started causing alarm

Photograph taken in august 1910, when subsoil grouting started.

Figure 23. Photographs during the construction of the Palace of Fine Arts, 1906 and 1910.

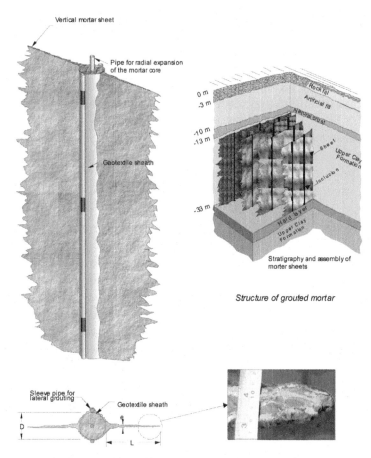

Figure 24. Rigid inclusions with lateral sheets of mortar.

173

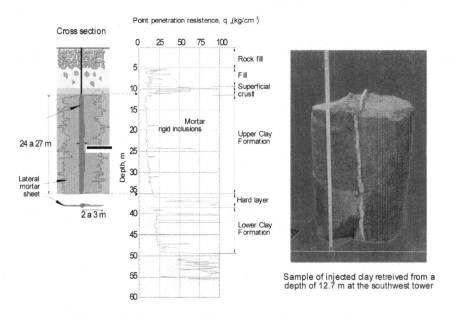

Point penetration resistence, q (kg/cm²)

Cross section

Sample of injected clay retreived from a depth of 12.7 m at the southwest tower

Figure 25.   CPT sounding to define inclusion depth and sample showing a mortar sheet.

11   SUBSOIL HARDENING

Construction of the Cathedral and the Sagrario Church ended in 1813 and settlements resulting from the consolidation of the clays because of the surcharges applied by both churches must have finished some 5 to 10 years afterwards. This is why it is certain that ongoing settlements and those in the future are being and will be induced by the depletion of pore water pressure due to deep water extraction. The prognosis of long-term hydraulic conditions at the Cathedral, depicted in Fig. 6, justifies the decision for hardening differentially the subsoil under the Cathedral and the Sagrario, with the purpose of reducing accumulation of differential settlements in the future and improving the behavior of the structures.

Clay hardening at a modern building. Soil injection trials were carried out at the Government Building of the FES Zaragoza university campus, in order to show that the deformability of a rather large mass of Mexico City Clay could be reduced by creating within it the mortar structure described previously. Details of those trials exceed the scope of this document but the results of those trials were most satisfactory.

On site clay hardening tests. This test took place between November, 1997 and January, 1998 out side the southwest corner of the Cathedral; 18 rigid inclusions were built and 179.5 m³ of mortar were injected. The test site was monitored in detail by: a) measuring the evolution of settlements with a grid of control points;) recording the evolution of pore pressures with open head and electric piezometers before, during and after the test; c) measuring wave propagation velocities determined with a seismic cone as well as lateral stresses with a Marchetti dilatometer. These measurements went on until July 1998 and showed that settlements in the test zone reduced; hence, the test was considered to be a success.

Zones and percentages of grouting. The mortar grout was made with controlled amounts of cement, bentonite, pumice sand and additives. Reductions of deformability depend on the stiffness of the mortar and on the percentage of grout injected. This last concept is the ratio between the volume of mortar and the volume of soil to be improved, Figure 26. The borings to carry out the injection need to cross the thickness of the rock fill, of the archaeological fills and of the superficial crust and should then go through the clays of the Upper Clay Formation that were grouted down to their contact with the First Hard Layer.

Drilling techniques. In view of the complexity of the structures and their foundations, it was necessary to use several drilling equipments. Pneumatic and electric drilling rigs were adapted to be used within the narrow aisles of the crypts. Some of them were mounted on mobile bases to facilitate their transportation. Referring to the drilling tools, pneumatic bottom hammers, tri-cone bits, simple drag bits, and reamers were used at the depths where the holes would be grouted. In drilling from the atrium, heavy rigs mounted on vehicles were used; ski-mounted equipment was only utilized at certain stretches. Perforations were dug with a procedure similar to that used at the crypts, but with a somewhat larger diameter.

First soil hardening stage. The initial injection stage took place between September 1998 and September 1999 in the areas indicated in Figure 26, with grout percentages varying from 2 to 7% at the Cathedral and from 1 to 5% at the Sagrario. The inclusions were introduced in the Upper Clay Formation as follows: 419 at the Cathedral, 111 at the Sagrario, and 55 at the museum building in the northwest portion of the site. The Cathedral's southwest corner and the northeast south east corners of the Sagrario were grouted form 8 September 1998 to 4 June 1999. The Cathedral's southwest corner was injected in two stages, applying about 50% of the total amount in each. Later, from 7 June 1999 to 9 September 1999, the south zone received an injection of 2%. Distribution of mortars was decided taking into account the compressibility based zoning of the subsoil and the Observational Method.

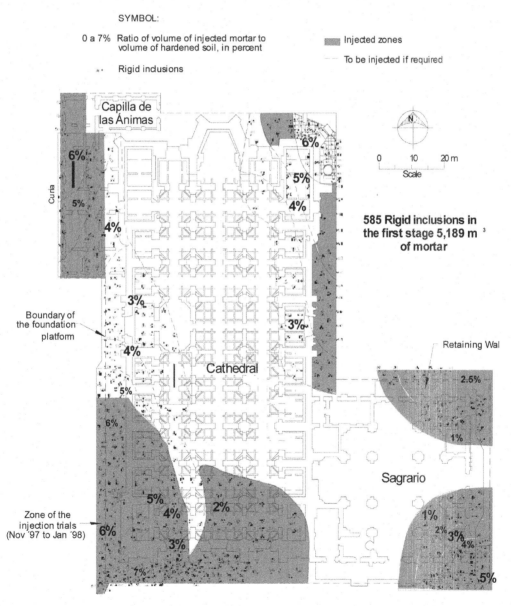

Figure 26. Distribution of mortars injected at the Cathedral. First stage: September 1998 to September 1999; second stage: May to July 2000.

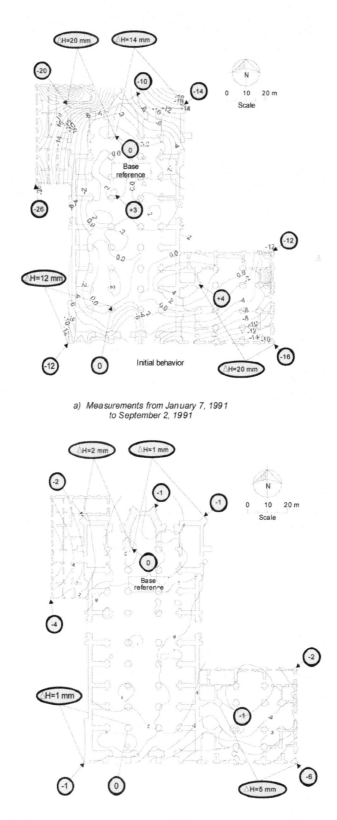

a) Measurements from January 7, 1991
to September 2, 1991

b) Measurements from 29 May 2009 to 24 October 2012

Figure 27.   Settlement contour rates in mm/year at the beginning of the project, 1991, and in 2012.

176

Second soil hardening stage. It took place from May to June 2000 and was complemented with injections that were carried out between 9 November and 22 December that year and with those made in the museum from 2 November 2001 to 20 January 2002.

A total of 585 inclusions were cast in the soft clays, together with their respective assemblages of lateral sheets. The total volume of injected mortar was 5,189 m³.

12 OBSERVED BEHAVIOUR

In applying the Observational Method, the structural behavior of both the Cathedral and the Sagrario church was monitored rigorously using a large number of measuring instruments. Soil response was evaluated mainly from high precision topographic surveys.

Efficacy of soil grouting for hardening purposes. The effectiveness of subsoil grouting can be evaluated by comparing settlement rates at the Cathedral and the Sagrario before and after injecting mortars. The left hand side of Figure 27 shows a picture of the initial behavior expressed graphically by plotting settlement rates observed between January 7 and September 2 of 1991 when the central part of the Cathedral emerged with respect to its northeast corner, at a rate of 16 mm/year, and with respect to the western bell Tower at 14 mm/year. The Sagrario shows a maximum settlement rate at its southeast corner of 16 mm/year with respect to point C-3, and between the southeast corner and the northwest columns, 20 mm/year.

The behavior in October 2012 shows that the Cathedral was sinking almost uniformly, as expected. From the direct comparison of the graphs in that figure, two conclusions can be derived: a) injection of mortar grouts into the subsoil modified positively the pattern of settlement rates; and b) this modification was beneficial for the structures because it achieved a substantial decrease of differential settlement rates. For example, relative differential settlement between the zero reference point and the southwest corner passed from 12 mm/year in 1989 to 1 mm/year in 20012. Overall, differential settlements in this last year were, on average, only a small portion of those existing in 1989, at the onset of the project.

13 FINAL COMMENTS

The goals in the project for the geometrical correction of the Cathedral and of the adjacent Sagrario church were established from previous experiences gathered while recovering the verticality of several other buildings with underexcavation. Furthermore, after adapting this method to a masonry structure, it was applied experimentally at the San Antonio Abad Church. Underexcavation at the Cathedral and the Sagrario started in August 1993. The preliminary goal was defined by Dr. Fernando López Carmona; subsequently, this goal was modified in 1994 by Dr. Roberto Meli. The geometrical correction achieved satisfies both proposals. Vertical corrective settlements after almost five years stabilized at a maximum of 88 cm.

The need to prevent the long-term effects of regional subsidence justified the application of mortar injections. The case history of the Teatro Nacional, now the Palace of Fine Arts, was taken as background information and reassessed in the light of present soil mechanics knowledge. The implementation of this method of subsoil hardening under the Cathedral was based on theoretical and experimental studies in the field and in the laboratory.

Recent evolution of differential settlements sustained by the Cathedral and the Sagrario has demonstrated that mortar injection of the subsoil has been a success given the beneficial effects it has had on the behavior of both churches. Settlement contours at the level of the plane of plinths confirm that historic settlement patterns were most favorably modified. The behaviors of both towers and of the reference plumb line corroborate the conclusions stated above.

This optimism follows from the comparison of the graphs presented in Fig. 13, but it shall have to be reconsidered in subsequent topographical and structural observations that could be even used to reach a better decision on whether subsoil hardening should be used in the future at the zones that remain untreated.

Data and analyses presented here demonstrate that subsoil hardening under the churches will be the alternative that, together with other complementary actions, will help preserve these monuments. Hardening has also the advantage of being a preventive method as opposed to underexcavation which is a corrective action. Recent measurements tend to justify the idea that soil hardening provides a long term solution to problems afflicting the Cathedral and the Sagrario church. Complementary adjustments can be made in the future to inject zones that may require to be treated in the years to come.

It should finally be acknowledged that the geometrical correction and subsoil hardening under the Cathedral and the Sagrario mitigate the harmful effects of differential settlements and are examples of projects meticulously controlled through the Observational Method. Results developed and applied there, when used in other cases will strengthen the capacity of engineers to face the risks associated to regional subsidence. This experience should lead the way and prompt practical research oriented towards the solution of many of the geotechnical problems still found in Mexico City.

Finally, it is important to state that this text offers a summary of geotechnical work carried out in the subsoil. Additional details into the geotechnical aspects of the project may be found elsewhere (Tamez et al 1997; Ovando-Shelley and Santoyo, 2001; Santoyo and Ovando, 2004).

ACKNOWLEDGEMENTS

Sergio Zaldívar, architect, headed the project since it began in 1989 until 2000; Dr. Xavier Cortés Rocha took over the direction of the project afterwards, up to 2009. The following members of the Technical Committee that overlooked the development of the project are duly acknowledged: Dr. Fernando López Carmona, Dr. Roberto Meli, Dr. Enrique Tamez, Hilario Prieto and Enrique Santoyo. Dr. Jorge Díaz Padilla acted as secretary for the committee and as consultants, Dr. Efraín Ovando- Shelley, Roberto Sánchez and Arturo Ramírez-Abraham.

REFERENCES

Carrillo, N. (1948), Influence of artesian wells in the sinking of Mexico City, *Proc. 2nd Int. Conf. on Soil Mech. and Found . Engng.,* Rotterdam, II.

Jamiolkowsky, M., Viggiani, C. and Burland, J. 1999. Geotechnical aspects. *Proc. Workshop on the Restoration of the Leaning Tower: Present Situation and Perspectives,* pre-prints Volume, Pisa.

Mazari, M., Marsal, R J and Alberro J. 1985. *The settlements of the Aztec great temple analyzed by soil mechanics,* Mexico: Mexican Society for Soil Mechanics.

Ovando-Shelley, E. and Manzanilla, L. 1997. Archaeological interpretation of geotechnical soundings in the Metropolitan Cathedral, Mexico City, *Archaeometry,* (39), 1,221–235.

Ovando-Shelley, E. and Santoyo, E. 2001. Underexcavation of buildings in Mexico City: the Case of the Metropolitan Cathedral and the Sagrario church. *Proc. ASCE, Journal of Architectural Engineering,* 7, (3), 61–70.

Santoyo, E., Ovando, E., Guzmán, X., Cuanalo, O. and y De la Torre, O. (1998). *Palacio de Bellas Artes. Campañas de inyección del subsuelo,* Mexico City: TGC Geotecnia.

Santoyo, E. and y Ovando, E. 2003. Cement injection in Mexico City for levelling buildings. Chap 12 in: *Passado, presente e futuro dos edifícios da orla marítima de Santos.* Sao Paulo, Brazilan Geotechnical Society.

Santoyo, E. and Ovando-Shelley, E. 2004. Geotechnical considerations for hardening the subsoil in Mexico City's Metropolitan Cathedral". In: *Advances in Geotechnical Engineering: The Skempton Conference,* Vol. 2. London: Thomas Telford, 1155–1168.

Tamez, E., Santoyo, E. and y Ovando, E. (1997). Underexcavation of Mexico City's Metropolitan Catedral and Sagrario Church". *Proceedings of the Fourteenth International Conference on Soil Mechanics and Foundation Engineering, Special invited lecture,* Volume 4. 2105–2126, Hamburg.

Terracina, F. (1962), Foundations of the Tower of Pisa, *Geotechnique,* London, 12, (3), 336–339.

Geotechnics and Heritage – Bilotta, Flora, Lirer & Viggiani (eds)
© 2013 Taylor & Francis Group, London, ISBN 978-1-138-00054-4

The Buen Pastor Cathedral in San Sebastián

A. Gens, E.E. Alonso & J. Casanovas
Technical University of Catalonia (UPC), Barcelona, Spain

L. Uzcanga
Architect, San Sebastian, Spain

ABSTRACT: The construction of an underground car park required an excavation in sand under the water table adjacent to the cathedral of Buen Pastor in San Sebastián, Spain. Because of the close proximity of the cathedral and other buildings, control of ground movements was a paramount consideration in the design. Construction involved the use of diaphragm walls propped by the floor slabs using the top-down technique. A monitoring system was installed to control the performance of the excavation throughout. For water control during the excavation, the diaphragm walls reached an underlying low permeability silt layer. This decision required the adoption of special measures to connect hydraulically the foundation ground of the cathedral with the water levels prevailing in the outside area. Excavation was completed with very limited ground movements that caused no observable damage in the cathedral and other nearby structures. Piezometer records also proved that the installed hydraulic connection performed satisfactorily.

1 INTRODUCTION AND HISTORICAL BACKGROUND

The Buen Pastor (Good Shepherd in English, Artzain Onaren in Basque) Cathedral is located in the city of San Sebastián (Donostia) in the Basque Country, Spain (Fig. 1). It was part of the 19th century expansion of the town outside the former old town precinct. Construction was initiated in 1888 in an area of marshy ground and sand dunes provided by the City Council, about 500 m from the current sea shore. Four different designs were submitted to the project competition; the local architect Manuel de Echave was declared the winner. He was also charged with directing the works. The church was opened for worship in 1897 although the spire was not completed until 1899 under a different architect (Ramón Cortázar). The planned total budget was 750,000 pesetas although the final cost was somewhat over twice that amount. Historical records of the construction indicated that the ground had to be drained and conditioned before proper building work could be started. The same records also proudly state that "all artisans involved were of Basque descent, highly competent, skilled and clever".

The cathedral covers an area of 1,915 m² with a Latin cross plan with a nave, a transept and a pentagonal apse (Fig. 2). The western nave has three aisles and the eastern nave and the transept five. The church has an overall length of 90 m of which 64 m correspond to the nave. The width of the western nave is 24 m whereas the width of the transept and eastern nave is 36 m. The maximum height of the church (central aisle) is 25 m and of the spire is 75 m. Sandstone from local quarries was employed for most of the construction but the vaults were built of tuff from the region whereas slate from Angers in France was used for the roof (Murugarren, 1996). The style was neo-Gothic, popular at the time, clearly inspired in French and German medieval examples. Architectural critics have identified significant influences of the Cologne cathedral, especially with respect to the spire.

In the nineties (about 100 years after construction), it was decided to construct an underground car park surrounding the cathedral on all four sides (Fig. 3). Due to space requirements, the distance between the excavation and the cathedral was very small in many locations. Design and construction procedures were selected to minimize ground movements that could affect not only the church

Figure 1.   The Buen Pastor cathedral.

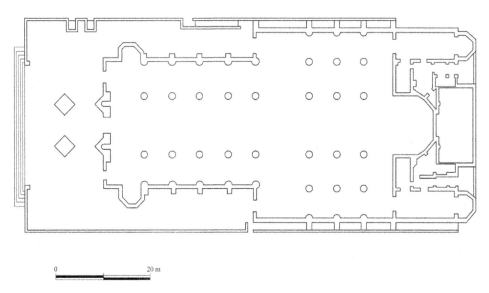

0   20 m

Figure 2.   Plan view of the Buen Pastor cathedral.

but other adjacent buildings as well. In this contribution the following aspects of the project and the construction are described and discussed: ground profile, design, excavation procedure, monitoring system and ground displacements. Hydraulic conditions and their impact on the works are also given due consideration.

180

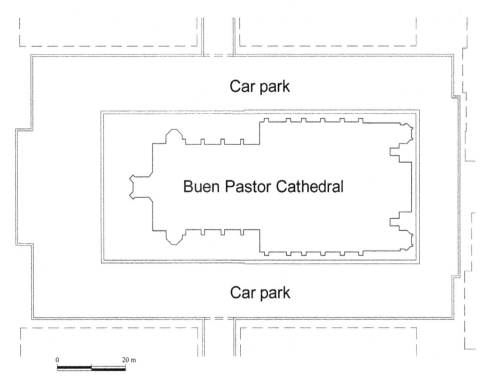

Figure 3. Layout of the planned car park surrounding the cathedral on all four sides.

2 GEOTECHNICAL CHARACTERISTICS OF THE SITE

2.1 *Ground profile*

The ground profile was determined from two site investigation campaigns. The first one was a preliminary exploration with 8 boreholes of limited length that failed to provide a complete picture of the underlying ground. The second one was more comprehensive; it included: five 42 m-long boreholes, five CPT tests and two pumping tests. The location of the boreholes and in situ tests are shown in Figure 4. From the site investigation, it was possible to identify the following mean ground profile:

– From 0 to 6 m depth. Fill containing sand and gravel size particles with boulders and, occasionally, construction debris. It is relatively well compacted with SPT values between 20 and 40.
– From 6 to 22 m depth. Poorly graded fine to medium sand. Most SPT values were in the range of 20–50 although, occasionally, higher values were obtained. Cone resistances lie in the 4–16 MPa range. Only in one zone the upper part of the layer exhibited a higher silt content and some organic matter, corresponding to the marshy conditions in some areas reported in the original construction records. Elsewhere, fine content was generally less than 4%.
– From 22 m to 35–40 m. Low plasticity soft silt (w_L = 32–38; I_p = 4–9). Generally, the proportion of fines was above 80% but, on occasions, fine content dropped to 20%–30%, the material becoming a silty sand. Oedometer tests gave values of compression index, C_c, of 0.13–0.23 in the more silty materials. Unconfined compression tests yielded very low values of undrained shear strength (16–25 kPa), unrealistic values given the depths of sample recovery, due, most probably, to sample disturbance in the extraction of this silty soil. CPT tests provided values of cone resistance consistent with a normally consolidated state of the layer.

Below the soft silt, bedrock appears that, in this area, is constituted by a flysch formation with closely interbedded shale and limestone layers. Due to the depth at which it appears, it has no influence on the project. Because it was important to the project, the thickness of the silt layer was ascertained carefully; it was consistently more than 6 m in all zones explored.

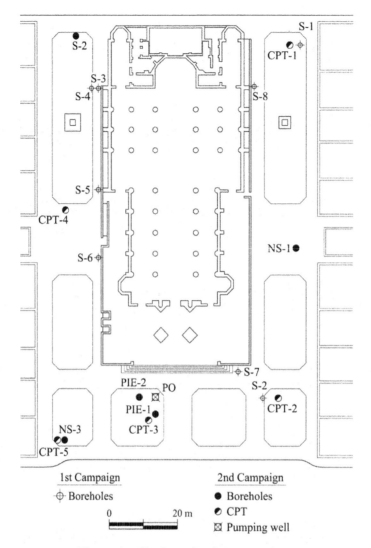

Figure 4. Site investigation campaigns.

Historical records indicated that the foundation was on (literally) "a layer of compressed sand". The footings were made of limestone masonry joined by hydraulic cement with a thickness varying between 1 and 3 m. Some short boreholes were drilled during the site investigation to verify this information. It was confirmed that the cathedral has a shallow foundation at a depth of 6 m that rests on the sand layer below the fill. It must be added that, at the moment of initiating the works, no damage to the building that could be attributed to ground settlements were observed. In fact, there was a crack on the floor of the cathedral running along most of its length. But this crack was not structural: it concerned only the thin floor slab placed directly on the fill.

2.2 Hydraulic conditions

Observations of piezometers placed in two boreholes were performed over a long period of time while the project was being developed. The mean water level depth was about 5 m, generally just above the contact between fill and the sand layer. Tidal variations of about 20 cm were consistently observed. Seasonal variations were larger but they never exceeded 1 m.

The permeability of the sand layer was determined by pumping tests that yielded a rather narrow range of results: 1.3–$2.5 \cdot 10^{-2}$ cm/s. The permeability of the soft silt was estimated from variable head permeability tests performed in the laboratory. The range here was wider, $2 \cdot 10^{-4}$–10^{-7} cm/s, depending on the fines content of the sample (Fig. 5).

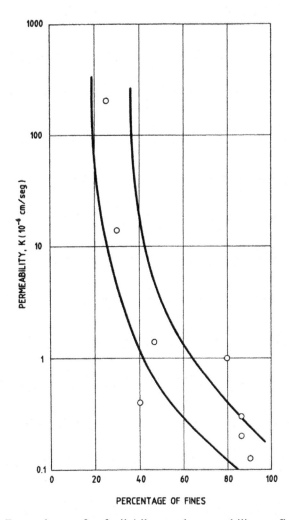

Figure 5. Dependence of soft silt/silty sand permeability on fines content.

3 MAIN FEATURES OF THE PROJECT

3.1 *General*

Obviously, the selection of the construction procedure was mainly controlled by the need of minimizing ground movements during excavation and construction. As indicated in Figure 3, the distance between the required excavation to the transept, eastern nave and the apse area is only about 1.5 m. Distances to the excavation are larger (5–7 m) around the rest of the cathedral. There was also the need to ensure no significant damage to the buildings on the opposite side of the works; they are located at distances of 2–3 m from the excavation boundary. Most of those buildings are old (a number of them with some historical significance as well) with load bearing walls that are potentially sensitive to ground movements.

The minimization of ground movements led to the selection of an excavation protected by diaphragm walls and frequent propping as the excavation proceeded. An efficient way to achieve this is by using top-down construction where the various floor slabs are constructed to serve as props (Fig. 6). This requires that the pillars of the structure are constructed prior to the excavation. Three car park levels were envisaged requiring a total excavation of 9.5 m, about 3.5 m below the foundation level of the cathedral. Figure 7 shows a more detailed sketch of the solution adopted where the two intermediate and the final excavation levels as well as the soil profile are indicated. The intermediate excavation levels were chosen to allow an easy construction of the corresponding floor slab and to obtain similar maximum bending moments in the diaphragm walls in all excavation stages. The required thickness for the

183

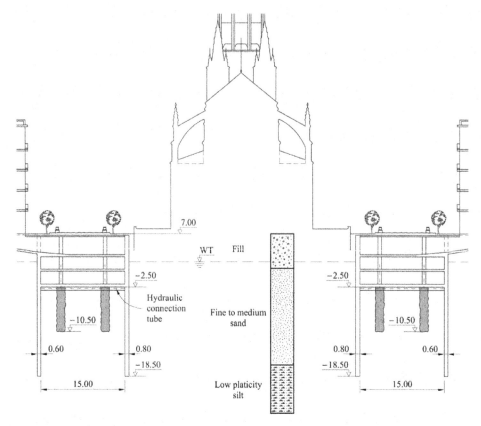

Figure 6. Adopted solution.

diaphragm walls was of 0.80 m for the wall immediately adjacent to the cathedral and 0.60 m for the wall on the opposite side of the excavation.

In order to reduce even further ground movements, consideration was given to the possibility to install a prop (using jet grouting or other techniques) below the bottom level of the excavation, prior to the start of works (e.g. Karlsrud & Andresen 2007, Shirlaw et al. 2006). However, numerical analyses indicated that, even without this additional prop, movements would be only of the order of millimeters, so this measure was not adopted. In one area (in the western part of excavation) temporary steel props and some ground anchors were used. This area was left temporarily free of floor slabs to provide an opening for extracting the excavated materials.

The construction of the pillars for top-down construction is easier if they are founded on single circular piles as precise control of position and verticality becomes simpler (Fig. 8). In this case, it was further decided to limit the length of the piles in order to keep a sufficient distance between the pile end and the soft silt layer. This was achieved by using 9 m long piles of 1.25 m diameter. They are totally embedded in the sand layer and are capable of supporting the loads of the structure with an adequate factor of safety.

3.2 *Water control*

As in any excavation in permeable material inside diaphragm walls, there are two possible choices: i) leave the toe of the wall in the permeable material and perform the excavation after water table lowering or after providing an impervious precinct by creating a low permeability base (using, for instance, some type of grouting), or ii) prolong the diaphragm walls until reaching an impervious layer. In this instance, choice ii) was clearly preferable. Creating a low permeability base was much more expensive than the alternative and lowering the water table was considered risky because of the likely consolidation settlements, especially in the soft normally consolidated silt layer. Therefore, the walls are significantly longer than required for structural reasons, but the solution is cheaper and safer.

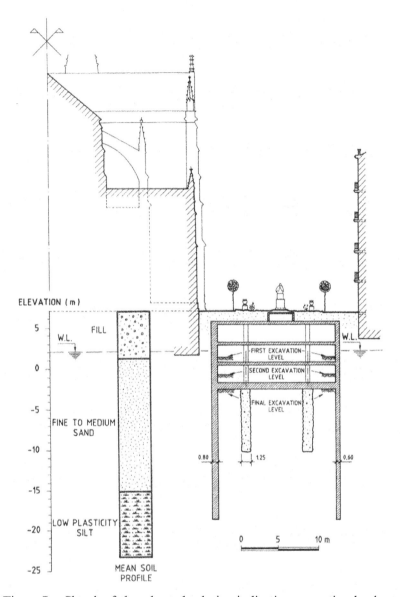

Figure 7. Sketch of the adopted solution indicating excavation levels.

Figure 8. Pillar embedded in a foundation pile (after excavation).

185

Diaphragm walls were excavated using bentonite muds for support. An important risk of excavations in sands under the water table using this procedure concerns the possibility of defects in the diaphragm walls concrete or, more frequently, in the joints between wall panels. Any such defect may lead to entry of water into the excavation that, if unchecked, can bring about sand migration and the production of uncontrolled large subsidence troughs. Accordingly, strict bentonite desanding and concreting procedures were specified as well as a stringent control of the location and verticality of the panels.

An unavoidable consequence of the selected excavation procedure is that the ground under the cathedral becomes hydraulically isolated. This is undesirable. If the diaphragm walls are perfectly tight the water level below the cathedral will rise continuously due to infiltration from the surface and other sources. If, in contrast, the walls are not watertight, the water level under the church may drop significantly leading to the development of settlements with time. Therefore, it is very important to maintain the water level in the inner enclosure in hydraulic equilibrium with the outer water level. During excavation this was achieved by leaving a gap in two opposite diaphragm walls protected by two plastic bentonite-cement walls that could be easily excavated at the end of the works. The permanent hydraulic connection required especial measures that are described at the end.

3.3 *Hydraulic uplift*

The final depth of the diaphragm wall was partially controlled by the need to avoid any potential hydraulic uplift of the bottom of the excavation. The variability of the permeability in the silt layer made it necessary to consider the possibility that a higher-permeability horizontal layer (associated with a lower fines' content) may be present at any depth. This criterion leads to a lengthening of the diaphragm walls until an adequate factor of safety against hydraulic uplift is achieved. A similar criterion was adopted for the construction of the Westminster car park (Burland & Hancock 1977). An additional criterion was ensuring that the walls were embedded a sufficient length in the silt layer to guarantee that the water entry into the excavation would be small.

Meeting these two criteria led to a penetration of 3.5 m in the silty layer; the total length of the diaphragm walls was 25.5 m. During the excavation of the panels, the material extracted was closely monitored to make sure that the prescribed minimum embedment in the impervious layer was achieved. The section of the walls without a structural function was left with a very light reinforcement.

4 MONITORING, EXCAVATION AND OBSERVATIONS

4.1 *Monitoring system*

An integral part of the project was the monitoring system for collecting relevant data before, during and after excavation. In addition to conventional surveying measurements of ground movements of the surface and of the adjacent buildings, piezometers, clinometers, inclinometers and sliding micrometers (Kovari & Amstadt, 1983) were installed. Inclinometers were placed in the ground adjacent to the diaphragm walls and the sliding micrometers were installed at different distances from the excavation. Piezometers were placed both inside and outside the excavation. Standpipes were used in the sand and vibrating wire piezometers in the silt. The locations of the various instruments are indicated in Figure 9. Instead of distributing the monitoring devices uniformly, most of them were concentrated in a few sections in order to check not only that ground movements were as small as expected but also to establish that the overall behaviour of the excavation and diaphragm walls was properly understood. The most comprehensively instrumented section is that denoted as P-P' in Figure 9, its cross-section is depicted in Figure 10. Incidentally, the locations of the cement-bentonite walls to ensure hydraulic equilibrium between the outside water table and the inner enclosure are also shown in Figure 9.

186

4.2 *Excavation performance and observations*

The first activity was the construction of the piles and the installation of the corresponding pillars. Afterwards, the top slab was concreted, an efficient measure to reduce future ground movements. Then, and in accordance with the top-down construction procedure, each excavation stage was accomplished along the full length of the works followed by the construction of the corresponding floor slab to provide the necessary propping action. Within each excavation stage, the sequence followed the numbering indicated in Figure 9. This meant that the information from the fully instrumented section was available early in each excavation stage. In addition, with this sequence, excavation was executed first in the zones further away from the cathedral; the excavation in the zone with

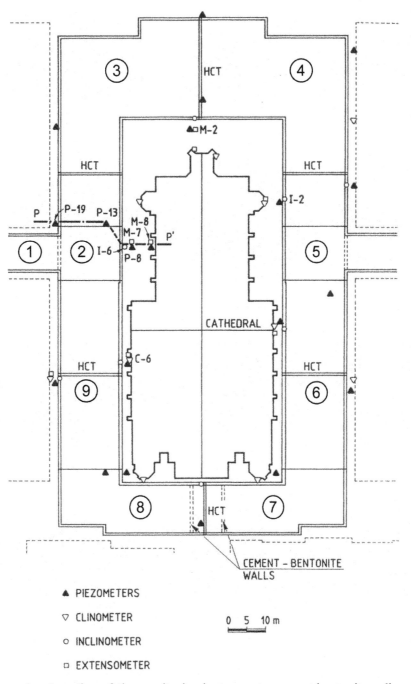

Figure 9.   Location of the monitoring instruments, cement-bentonite walls and hydraulic connection tubes (HCT).

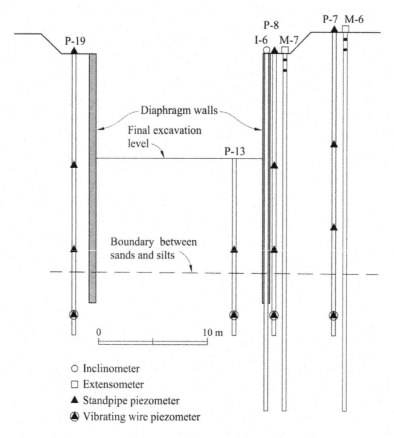

Figure 10. Instrumentation layout. Section P-P'.

a minimum distance to the cathedral was performed later. Diaphragm walls were closely watched as excavation proceeded, no significant defects were encountered and the water entry into the excavation through walls and panel joints was negligible. Figures 11 and 12 show views of the construction activities.

Generally, ground movements were small, typical values were about 3 mm for settlements and 0.5 mm for the lateral movements of the wall. The distributions of observed ground movements at the end of excavation for section P-P' is shown in Figure 13. Considering all observations, the maximum recorded settlement was 6 mm in micrometer M-2 and the maximum lateral movement was 1.5 mm in inclinometer I-2. Figure 14 shows those maximum displacements compared with other cases collected in Clough et al. (1990). They plot well below the average values reported. A more recent collection that distinguishes between different support options has been presented by Long (2001). It can be again be noted (Fig. 15) that the displacements caused by the excavation lie in the lowest values range, no matter which support system is considered. Readings of clinometers placed on adjacent buildings and on the cathedral were also very small. The maximum value (0.69 10^{-3}) was recorded in clinometer C-6 (Fig. 9) but, at the end of the works, it had recovered much of this inclination.

A typical time evolution of piezometer readings before, during and after excavation is presented in Figure 16. They also correspond to the fully instrumented section P-P'. It can be observed that the piezometric level inside the excavation (P-8) is the same as the piezometric level outside the excavation (P-19) confirming the effective hydraulic connection between the two zones throughout excavation and beyond. Also, no water level changes outside the excavation is measured. The evolution of the readings of piezometer P-13, located in the sand inside the excavation, reflects the lowering of the water table inside the walls as excavation deepens. In contrast, the vibrating wire piezometer placed also inside the excavation but in the silt layer, responds slowly to the water level changes in the sand.

Figure 11.   Overall view of construction activities.

Figure 12.   Excavation in progress. The figure shows the boundary zone between the excavation propped with floor slabs and the excavation supported by ground anchors.

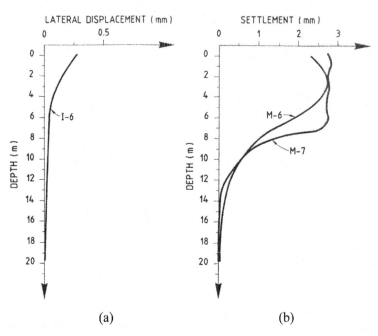

(a)   (b)

Figure 13. (a) Lateral wall displacements in inclinometer I-6. (b) Distribution of vertical settlements in micrometers M-6 and M-7.

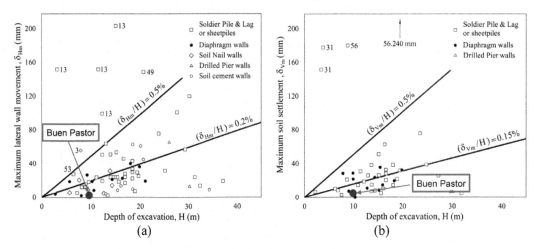

(a)   (b)

Figure 14. Comparison of measured maximum displacements with previous results reported in Clough et al. (1990). (a) Maximum lateral wall displacements. (b) Maximum settlements.

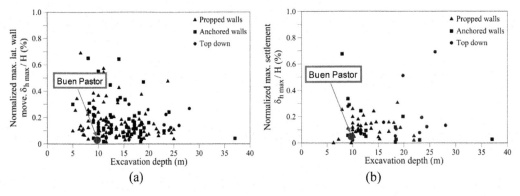

(a)   (b)

Figure 15. Comparison of measured maximum displacements with previous results reported in Long (2001). (a) Maximum lateral wall displacements. (b) Maximum settlements.

190

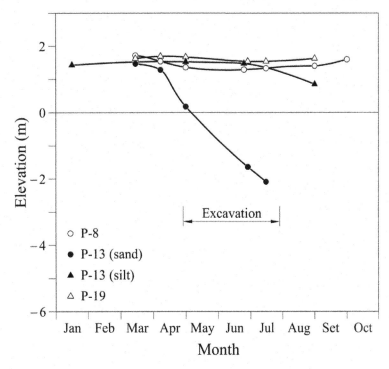

Figure 16.   Evolution of piezometric levels in section P-P′.

As a consequence, settlements caused by this reduction of pore pressure are very limited. Naturally, pore pressures in the silt layers returned to the initial values once the impervious bottom slab was constructed.

5   PERMANENT HYDRAULIC CONNECTION

At the end of construction, the cement bentonite walls above the bottom excavation level were dismantled and the temporary hydraulic connection was closed. To ensure a permanent hydraulic connection, a series of tubes connecting the outside ground with the inner enclosure were constructed just below the maximum excavation level. Calculations demonstrated that six connection tubes were amply sufficient to ensure hydraulic equilibrium. Their location in plan is shown in Figure 9 and in a cross-section in Figure 17. Each end of the tubes was protected by a geotextile filter placed on a porous concrete plug.

Each tube linked two opposite cylindrical holes that had been left at the required locations in the diaphragm walls. The cylindrical holes were temporarily closed with a sealed steel plate. After installing the connecting tube, the contact between the wall concrete and the tube was hydraulically sealed. To prevent an inrush of water and sand when withdrawing the steel plate and placing the tube, the ground adjacent to the opening was frozen using liquid nitrogen. Once the tube was installed and sealed, freezing was discontinued allowing water flow to take place. As an additional measure to increase the flow towards the hydraulic connection zones, a vertical borehole was drilled and equipped with a perforated drainage tube protected by a geotextile. A detailed scheme of the tube ends is shown in Figure 18 whereas Figure 19 contains some pictures of the hydraulic connection tubes. This connecting system proved to be quite successful; no water level differences have been measured between the general water table outside and the water level under the cathedral.

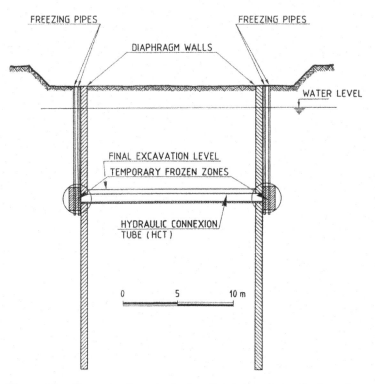

Figure 17.   Cross-section at a hydraulic connection tube location.

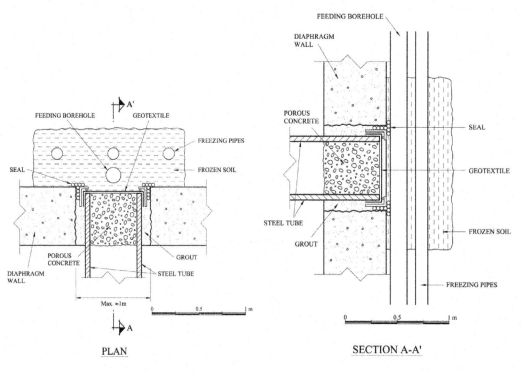

PLAN

SECTION A-A'

Figure 18.   Layout of the ends of the hydraulic connection tubes.

<table>
<tr><td>(a)</td><td>(b)</td></tr>
</table>

Figure 19. (a) Prepared seat for the hydraulic connection tube. (b) Porous concrete plug at one end of the hydraulic connection tube. The protective geotextile not installed yet.

6 CONCLUDING REMARKS

An excavation for the construction of an underground car park has been performed in sand under the water table very close to the Buen Pastor Cathedral and other buildings. The distance of the excavation to the cathedral was of the order of only 1.5 m in many areas. It can be concluded that, with a good selection of the construction procedure and a well-controlled and closely monitored construction, it is possible to limit ground movements, and hence damage to adjacent structures, to very small magnitudes. Because the solution adopted involved the hydraulic isolation of the foundation ground of the cathedral, it was necessary to undertake special measures to ensure the hydraulic connection of the inner enclosure with the water level prevailing around the area.

REFERENCES

Burland, J.B. & Hancock, R.J.R. 1977. Underground car park at the House of Commons, London: geotechnical aspects. Structural Engineer, 55(2): 87–105.

Clough, G.W. & O'Rourke, T.D. 1990. Construction induced movements of in situ walls. In Lambe P.C. & Hansen L.A., (eds.), Proc. Design and Performance of Earth Retaining Structures, ASCE Special Technical Publication 25: 439–470.

Karlsrud, K. & Andresen, L. 2007. Design of deep excavations in soft clays. In Cuéllar et al. (eds.), *Proc. 14th European Conference on Soil Mechanics and Foundation Engineering, Madrid*, Rotterdam: Mill Press 1: 75–99.

Kovari, K. & Amstad, Ch. 1983. Fundamental of deformation measurements in Geomechanics. Proc. Int. Symp. on Field Measurements in Geomechanics, Zurich, 1: 29–239.

Long, M. 2001. Database for retaining wall and ground movements due to deep excavations. *J. Geotechnical and Geoenv. Eng.*, ASCE, 127: 203–224.

Murugarren, L. 1996. *Catedral del Buen Pastor. Donostia-San Sebastián, 1897–1997*. San Sebastián: Kutxa Social and Cultural Foundation.

Shirlaw, J.N., Tan, T.S. & Wong, K.S. 2006. *Deep excavations in singapore marine clay. In Bakker et al. (eds.), Geotechnical Aspects of Underground Construction in Soft Ground, London: Taylor & Francis Group, 13–28.*

Geotechnics and Heritage – Bilotta, Flora, Lirer & Viggiani (eds)
© 2013 Taylor & Francis Group, London, ISBN 978-1-138-00054-4

Stability problems of leaning towers on weak soil

R. Lancellotta
Politecnico di Torino, Torino, Italy

ABSTRACT: The fascination of a leaning tower actually hides problems related to the interaction with the foundation soil, sometime not so evident at first glance. In particular, initial imperfections and non-uniform foundations very often generated differential settlements which, coupled with low stiffness of the soil-foundation system and delayed deformations, can led to long-term critical conditions.

A convenient way to forecast this long term behavior, as well as to study the effect of seismic action using spectral analysis, is to rely on the inverted pendulum model, provided we can define a reliable response for the soil-foundation system.

Strain-hardening plasticity models, developed in relation to the overall soil-footing system (i.e. the so-called "macro-element" model), can provide such a relationship, but, in any case, it is rather difficult to validate these models, by considering the 3D nature of the problem, as well as the effectiveness of the soil contact along the vertical sides of the embedded foundation.

In this respect, a rather significant contribution comes from an experimental identification analysis, performed in the presence of ambient vibration, as conceptually proved by the dynamic analysis of stability.

1 INTRODUCTION

Historic towers, in particular medieval towers, are an important part of our cultural heritage and their preservation claims for monitoring, for detailed analyses of their long term performance as well as of their vulnerability under seismic actions.

There are aspects of structural nature, linked to the masonry behaviour as a unilateral material, that deserve special attention as proved by the collapse of the Campanile in Venice and the Civic Tower in Pavia (Heyman, 1992; Binda et al., 1992; Macchi, 1993). And there are aspects related to the interaction with soft soil conditions that play a major role in determining the long term performance of the tower.

We have to remind that during the first stage of construction the tower could have been not so far from a *bearing capacity* collapse, due to *lack of strength* of the soil, and safely survived thanks to some delay or interruption of the building process, that allowed the foundation soil to improve its strength and the tower to be successfully finished.

Then, a more subtle and often unforeseen danger was represented by leaning instability, that increases if there is *lack of stiffness* of the soil. And, because of creep phenomena, a tower, even if initially stable, can attain with time an instability condition (*asymptotic leaning instability*).

Futhermore, it must be outlined the severe damage that occurred and may occur to monument and cultural heritage during seismic events. The seismic sequence of May 20 and 29, 2012, in Emilia Romagna (Italy), unfortunately offers in this respect recent and unexpected examples: damage occurred to the majority of monuments and there was the collapse of many campaniles, despite the moderate magnitude of the event ($M_L = 5.9$).

Moving from these considerations, this paper is intended to contribute to the preservation of historic towers by focusing on the role of soil-structure interaction on the long terms performance of the structure, i.e. by exploring the problem of the asymptotic leaning instability of tall towers on weak soil.

As expected, when exploring this problem, as well as when forecasting the effect of seismic actions using spectral analysis, a crucial step in the analysis is to establish a reliable response for the soil-foundation system and it will be shown that a novel and rather significant contribution comes from experimental identification analyses.

2 THE LEANING INSTABILITY

The fascination of a leaning tower actually hides problems related to the interaction with the foundation soil, sometime not so evident at first glance. At the same time we can notice how the observer, even if not aware of scientific knowledge, has the ability of capturing the essence of the problem, because, when asking himself about the reaction of the tower if perturbed by any external action, he is just disputing about the *stability of equilibrium*. The danger of a *leaning instability* increases if there is *lack of stiffness* of the soil, and in this respect the leaning tower of Pisa represents a powerful example, because the preservation of the tower was recognised as being a problem of leaning instability (Hambley, 1985; Lancellotta, 1993; Como, 1993; Desideri and Viggiani, 1994; Nova and Montrasio, 1995; Federico and Ferlisi, 1999; Burland et al., 2003; Marchi et al., 2011).

Modeling tall towers is a challenging problem, its complexity arising from the interaction between a number of geotechnical and structural phenomena. Initial imperfections and non-uniform foundations very often generated differential settlements which, coupled with low stiffness of the soil-foundation system and delayed deformations, can led to their long-term critical conditions.

A convenient way to explore the leaning instability mechanism is to make reference to the inverted pendulum model (Figure 1). The tower is represented by a rigid column of length h, connected at its base to a rotational spring, representing the soil-foundation system, and a crucial step in the analysis is therefore to establish a reliable response for the tower foundation.

To this aim, strain-hardening plasticity models, developed in relation to the overall soil-footing system (i.e. the so-called "macro-element" model), can provide such a relationship (Prager, 1955; Butterfield, 1980; Desideri and Viggiani, 1994; Nova and Montrasio, 1991; Paolucci et al., 1999; Como, 2010; Marchi et al., 2011), and allow to take into account the main aspects of soil behaviour: nonlinearity, irreversibility and coupling between settlement and rotation.

Within the framework of a hardening plasticity soil behaviour, the first step is to assume a yield function f which depends on the state of the so-called *generalized stress variables* (the overturning moment M and the component N of the weight P normal to the foundation plane) and on a set of hidden variables, which are in turn function of the irreversible *generalized strains* (the rotation α and the settlement w, that identify the motion of the tower).

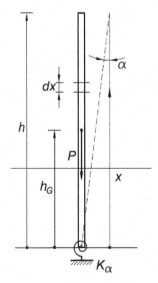

Figure 1.   The inverted pendulum model.

Let assume the yield function to be of the same form as the failure locus (Figure 2) suggested by Meyerhof (1953)

$$f = M - \frac{Na}{2}\left(1 - \frac{N}{N_{max}}\right), \tag{1}$$

where a is the side dimension of a squared foundation and N_{max} is the limit load in absence of any eccentricity.

A geometrical meaning can be associated to the equation (1) if we assume that the generalised stress component (N, M) are coordinates of an abstract space, since in this space the yield function is now representing a yield surface (Figure 2).

For a stress state inside this surface only elastic deformations occur and the yield surface can be so far interpreted as the instantaneous boundary of the elastic domain, the adjective instantaneous being included to recall that the yield surface evolves, acconding to a hardening parameter N_c, that depends only on irreversible strains.

Plastic increments of generalized strains develop if both the following conditions are satisfied

$$f = 0 \tag{2}$$
$$df = 0 \tag{3}$$

i.e. plastic deformations develop if the point, representative of the current stress state, lies on the yield surface and the stress increment is such that the point moves on it.

In order to derive the increments of plastic deformation a scalar function $g = g\,(N,M)$ is introduced, that plays the role of a *plastic potential*, so that

$$d\alpha = \Lambda\frac{\partial g}{\partial M}, \tag{4}$$
$$dw = \Lambda\frac{\partial g}{\partial N}$$

where Λ is a positive scalar.

Equation (3), known as *consistency condition*, allows to derive the plastic multiplier Λ and by substituting into (4) the following expression are obtained for the increment of plastic rotation and settlement

$$d\alpha = \frac{1}{H}\left(\frac{\partial f}{\partial M}\frac{\partial g}{\partial M}dM + \frac{\partial f}{\partial N}\frac{\partial g}{\partial M}dN\right) \tag{5}$$

$$dw = \frac{1}{H}\left(\frac{\partial f}{\partial N}\frac{\partial g}{\partial N}dN + \frac{\partial f}{\partial M}\frac{\partial g}{\partial N}dM\right), \tag{6}$$

where H is the hardening modulus.

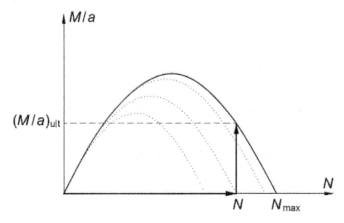

Figure 2.   Yield surfaces and failure locus.

Note that, in contrast to constitutive relationships predicted by elasticity, these equations prove that there is coupling between the components of generalized stress and strain.

Since equation (1) implies

$$\frac{\partial f}{\partial M} = 1 \quad \text{and} \quad \frac{\partial f}{\partial N} = a\left(\frac{N}{N_c} - \frac{1}{2}\right),$$

assuming associated flow rule ($f \equiv g$) and by considering a constant vertical load path (i.e. $dP = 0$) equation (5) allows to write

$$H = \frac{dM}{d\alpha},$$

so that the hardening modulus H assumes the meaning of tangent rotational stiffness.

If we postulate that such a stiffness progressively reduces from an initial value K_α^o (linked to the current values of the load N) to zero, depending on the distance of the current point (N, M) from its image (N, M_R) on the failure locus along the considered constant vertical load path ($dN = 0$), i.e.

$$H = K_\alpha^o\left(1 - \frac{M}{M_R}\right) \tag{7}$$

by substituting (7) into (5) we get

$$d\alpha = \frac{1}{K_\alpha^o}\frac{M_R}{M_R - M}dM \tag{8}$$

and the integration of (8) gives the solution

$$\alpha = \frac{M_R}{K_\alpha^o}\ln\left(\frac{M_R}{M_R - M}\right) \tag{9}$$

or the direct constitutive relationship (Como, 1993; Lancellotta, 1993; Federico and Ferlisi, 1999; Marchi et al., 2011)

$$M = M_R\left(1 - e^{-(\alpha-\alpha_o)/\gamma_R}\right) \tag{10}$$

where

$$\gamma_R = \frac{M_R}{K_\alpha^o} \tag{11}$$

and α_0 represents any initial rotation, due to imperfections of the system or uneven settlements during construction.

We outline that equation (10) implicitly depends on N, because both $M_R = M_R(N)$ and $\gamma_R = \gamma_R(N)$, and provides a first important insight into the basic mechanism of the leaning instability.

The equilibrium condition in the deformed configuration requires

$$Ph_G\sin\alpha = M_R\left[1 - e^{-(\alpha-\alpha_o)/\gamma_R}\right], \tag{12}$$

and if the initial tilt is zero and the constraint has a linear behaviour, by setting for small values of the current tilt

$$\sin\alpha \cong \alpha$$

$$e^{-\alpha/\gamma_R} \cong 1 - \alpha/\gamma_R,$$

198

the equilibrium is assured until a critical load is reached, this being given by

$$P_c = \frac{M_R / \gamma_R}{h_G} = \frac{K_\alpha}{h_G}.$$ (13)

When the load is equal to P_c, the possibility of having neutral equilibrium requires a restraint with constant stiffness K_α; on the contrary, in the presence of non-linear restraint any increase of tilt produces a sudden collapse, as shown by the curve labelled a in Figure 3.

If there is an initial tilt α_0, the current tilt increases for any given load P and, due to the combined effect of non-linear response and initial tilt, the collapse load is only a fraction of the critical load given by equation (13).

Rotational creep, unlike vertical creep under constant vertical load, affects the load state of the foundation and a tower, even if initially stable, can attain with time an instability condition.

This can be highlighted with refence to the plane (α, M) in Figure 4, where the external overturning moment $M_e = P h_G \sin \alpha$ and the resistant moment M_r are plotted (Cheney et al., 1991; Desideri and Viggiani, 1994; Marchi et al., 2011). If the initial slope K_α^0 of M_r is lower or equal to the slope (Ph_G) of the external moment load path M_e, then the M_r and M_e paths will never intersect and equilibrium

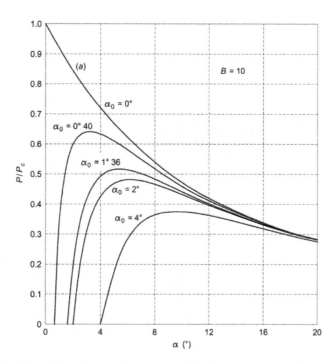

Figure 3.   Influence of soil nonlinearity on the post-peak behaviour of a leaning tower (Lancellotta, 1993).

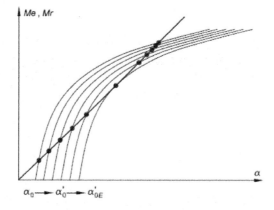

Figure 4.   Time evolution of the stability condition.

199

is never possible. By contrast, if $K_\alpha^o > Ph_G$, for an initial α_0 a stable equilibrium condition occurs at the point on the M_r curve where $dM_r/d\alpha > dM_e/d\alpha$.

The maximum value of M_e is attained when the M_r curve is tangent to the M_e line, and this may eventually occur if we assume that the effect of delayed deformation (creep) is that of increasing the initial tilt α_0. Therefore, in this simplified interpretation, the effect of the creep is to merely translate the M_r curve along the α axis, while its shape is assumed not to be affected by creep deformation (see Marchi et al, 2011, for a creep-hardening model).

In order to obtain a time evolution law of the rotation, we can assume, as first suggested by Como (1993), a viscous constitutive law based on the so called hereditary mechanics (Krall, 1947; Nadai, 1950; Como, 2010), that gives the material a permanent memory and reproduces many important aspects of the viscous behaviour of the foundations

$$\dot{\alpha}^v = \frac{M}{K_\alpha^v}\left[e^{-\beta(t-t_o)}\right].$$ (14)

The integration of (14) gives

$$\alpha^v = \frac{M}{\beta K_\alpha^v}\left[1 - e^{-\beta(t-t_o)}\right],$$ (15)

and therefore we can read the quantity $K_\alpha^v = M/\alpha_\infty^v$ as the rotational asymptotic viscous stiffness of the foundation and the positive constant β is a scale factor, dimensionally inverse to the time t (in the following century^{-1}).

The rotation rate at time t will be given by the plastic and viscous components (at this stage we are neglecting the elastic contribution)

$$\dot{\alpha} = \dot{\alpha}^p + \dot{\alpha}^v.$$

and the previous equations (8) and (15) allow to write

$$\dot{\alpha} = \frac{M_R}{K_{\alpha p}^o}\frac{\dot{M}}{M_R - M(t)} + \frac{M(t)}{K_\alpha^v}e^{-\beta(t-t_o)}.$$ (16)

As the rotation of the tower is certainly small, the dependence of N on the progressive tilting can be neglected, so that $N = P$, and the overturning moment can be expressed as

$$M(t) = Nh_G \sin\alpha(t).$$ (17)

Therefore

$$\dot{M}(t) = Nh_G \dot{\alpha}\cos\alpha(t)$$ (18)

and by substituting (18) into (16) and introducing the quantity $M_R/Nh_G = \mu_R$, one get

$$\dot{\alpha} = \frac{M_R}{K_{\alpha p}^o}\frac{\dot{\alpha}\cos\alpha(t)}{\mu_R - \sin\alpha(t)} + \frac{1}{K_\alpha^v}e^{-\beta(t-t_o)}Nh_G\sin\alpha(t).$$ (19)

Let now

a. define the quantity $\chi_\alpha^v = K_\alpha^v/Nh_G$;
b. introduce the quantity $\chi_\alpha^e = K_\alpha^e/Nh_G$ that takes into account the elastic component of the rotation by means of the elastic stiffness K_α^e and,
c. introduce a change of variables $\xi = e^{-\beta(t-t_o)}$,

then equation (19) can be integrated under the initial condition $\xi = 1$ (i.e. $\alpha = \alpha_0$) if $t = t_0$, to give the following evolution law (first derived by Como, 1993)

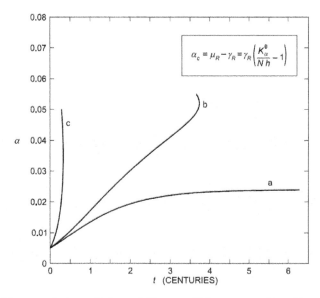

Figure 5. Time evolution of the stability conditions as predicted by equation 20.

$$\xi = 1 + \beta \chi_\alpha^v \left[\ln\left(\frac{\tan(\alpha_o/2)}{\tan(\alpha/2)} \right) + \frac{\gamma_R}{\mu_R} \ln\left(\frac{\mu_R - \sin\alpha_o}{\mu_R - \sin\alpha} \frac{\sin\alpha}{\sin\alpha_o} \right) + \frac{1}{\chi_\alpha^e} \ln\frac{\sin\alpha}{\sin\alpha_o} \right]. \tag{20}$$

The obtained evolution law proves that the tilt α increases with time (see Figure 5), but its rate $\dot{\alpha}$ can increase or progressively decrease with time depending on the elasto-plastic stiffness and on the asymptotic viscous stiffness, therefore we can reach a long term stable condition as well an instability condition.

The critical value attained by the rotation must correspond to $\frac{d\alpha}{dt} = \infty$ or $\frac{d\xi}{dt} = 0$, and this condition gives

$$\mu_R = \gamma_R \cos\alpha_c + \sin\alpha_c. \tag{21}$$

The rotation α being usually small, we can give to (21) the following form

$$\alpha_c = \gamma_R \left(\frac{K_\alpha^o}{Nh_G} - 1 \right), \tag{22}$$

that highlights the second order effects.

3   IDENTIFICATION ANALYSES

In order to reach quantitative conclusions, there are the following basic parameters to be assessed: the ultimate moment M_R (note that in equation (20) $\mu_R = M_R/Nh$), the initial stiffness K_α^o ($\gamma_R = M_R/K_\alpha^o$), equal to the elastic stiffness K_α^e or to the elasto-plastic one depending on whether the loading path is evolving inside or on the yield surface, and the so called asymptotic viscous stiffness βK_α^v ($\beta \chi_\alpha^v = \beta K_\alpha^v/Nh_G$).

We can certainly rely on strength parameters when assessing the ultimate moment and we can also rely on feed-back analyses to evaluate the asymptotic viscous stiffness. On the contrary, althoug the values of the rotational stiffness can been obtained by using the previously mentioned macro-element approach, it is rather difficult to validate the values so obtained, by considering the 3D nature of the problem, the mechanical heterogeneity induced by the applied stresses as well as the effectiveness of the soil contact along the vertical sides of the embedded foundation. And the complexity of the problem is such that this remark applies even if we consider an elastic behaviour.

In this respect, a rather significant contribution comes from experimental identification analyses, due to the relation that exists between the results of these procedures and the dynamic nature of the problem we are dealing with, as it will appear in the sequel.

As a first step, we have to recall that instability is a dynamic process and for this reason we have to approach the problem from a dynamic point of view, because this is a necessary step to define the concept of stability. This approach will also allow to explore the reason way we have previously analysed the stability problem by neglecting its dynamic nature.

To answer this question, let consider again the inverted pendulum model (Figure 1) and analyse the dynamic equilibrium of the tower perturbed under the initial conditions

$$\alpha = \alpha_o \qquad\qquad \dot{\alpha} = \dot{\alpha}_o.$$

If ρ is the mass per unit length, by taking into account the inertial forces, the dynamic equilibrium writes

$$Ph_G \sin\alpha - K_\alpha \alpha - \int_o^h (x\ddot{\alpha})\rho x\,dx = 0. \tag{23}$$

As usual $\sin\alpha \cong \alpha$, therefore we obtain

$$\ddot{\alpha} + \frac{3(K_\alpha - Ph_G)}{\rho h^3}\alpha = 0, \tag{24}$$

and we have to explore two cases.

a. If $Ph_G < K_\alpha$, by introducing the quantity

$$\omega^2 = \frac{3(K_\alpha - Ph_G)}{\rho h^3} > 0, \tag{25}$$

the solution attains the following form

$$\alpha(t) = \alpha_o \cos\omega t + \frac{1}{\omega}\dot{\alpha}_o \sin\omega t. \tag{26}$$

In this case, the solution is bonded and, if initial values are small, oscillations remain small for all times, i.e. perturbations or changes in the system produce small oscillations and the quantity ω represents the circular frequency of the system.

b. On the contrary, if $Ph_G > K_\alpha$,

$$\eta^2 = \frac{3(Ph_G - K_\alpha)}{\rho h^3} > 0, \tag{27}$$

the solution

$$\alpha = \frac{1}{2}\left(\alpha_o + \frac{\dot{\alpha}_o}{\eta}\right)e^{\eta t} + \frac{1}{2}\left(\alpha_o - \frac{\dot{\alpha}_o}{\eta}\right)e^{-\eta t} \tag{28}$$

now proves that an instability condition is reached, because no matter how small the initial values are, the rotation becomes increasingly large.

Therefore, since the state of the system depends on the parameter $(K_\alpha - Ph_G)$, that represents the effective stiffness, this parameter varies the state branches at a critical value (a bifurcation or branching is possible) with a change of stability, i.e. if the quantity $(K_\alpha - Ph_G)$ is positive the motion can be considered oscillatory with the frequency ω, whereas it degenerates if the effective stiffness vanishes.

Instability then occurs at $\omega = 0$, the motion is at constant velocity and inertial forces vanish, and this explains why the previous static analysis provides the same critical load.

This means that once the critical load $P_c = K_\alpha / h_G$ is attained, there is *lack of stiffness* with respect to a possible deformed configuration, that superimposes to the trivial one, giving rise to a bifurcation of paths.

In addition to these remarks, the dynamic analysis suggests in a rather natural way the link between the frequency of oscillation and the stiffness of the system, so that measurements of dynamic properties (frequencies, mode shape and damping) can provide the identification of the stiffness of the constraint and the deformability modulus of the masonry.

Dynamic tests can be performed under known excitation or unknown environmental excitation. In the first case the tests are performed using artificially produced excitation of harmonic type at variable frequency, usually by applying to the structure an eccentric mass rotating machine (vibrodine), or by producing a pulse load by means of a dropping weight.

Although they offer advantages, i.e. the different modes can be excited individually, modal shapes and damping can then directly be identified and non linear behaviour can be also explored, these tests involve several shortcomings. These include difficulties in setting up the testing equipment, the need to prevent the use of the structure during testing, resolution problems associated with closely frequency coupled modes, very long execution times. As a consequence, they are not suitable for continuous long-term monitoring projects.

In addition, the use of "ad hoc" excitation often proves very costly and sometimes difficult to put into practice, as in the case of old masonry structures or severely damaged buildings.

In contrast, dynamic tests with environmental excitation offer the following advantages: the use of the structure does not have to be interrupted, no special excitation equipment is required and excitation does not have to be known or measured. Test results can be continuously recorded on time, and it is possible to perform an unlimited number of tests without impairing the integrity of the structure.

The capability of these tests has been proved for the case of the Ghirlandina tower in Modena (Lancellotta, 2009; Sabia, 2012; Di Tommaso et al., 2013; Lancellotta, 2013) and is discussed in details in the Keynote Lecture by Lancellotta and Sabia (2013). The main conclusion reached in these analyses was that the soil structure interaction cannot be neglected, in contrast to most published identification analyses that usually assume the structure with rigid constraint at its base.

Furthermore, it must be outlined that the stiffness values obtained by identification analyses usually may be considered in many cases as representative of an elastic response, to be used with reference to serviceability limit states or in presence of low intensity seismic events.

One reason for such behaviour may be attribute to hardening due to creep, as it shown in Figure 6, where reference is made to a triaxial tests and the stress and strain parameters are the mean effective stress p', the deviator stress $q = \sigma_1 - \sigma_3$ and the deviator strain $\varepsilon = \frac{2}{3}(\varepsilon_1 - \varepsilon_3)$.

If a load path if performed from A to B and then a rest period takes place from B to C, due to delayed deformations the yield locus expand with time, so that during subsequent loading from C to D the material will behave elastically, whilst elasto-plastic behaviour will occur from D to failure (Nova, 1982; Marchi et al., 2011).

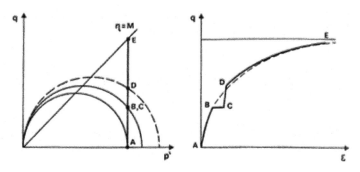

Figure 6.   Stiffnening of stress-strain response due to creep effects (Nova, 1982).

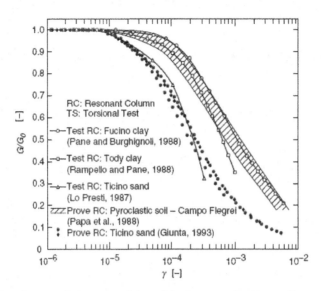

Figure 7. Decay of shear modulus with shear strain (Lancellotta, 2009.a).

In order to capture the elasto-plastic response, as well as when the interest is to describe the complete evolution of the tilt with time of towers on weak soils, we have to take into account the decay of shear modulus with shear strain (Figure 7) and to properly select the shear modulus according to the expected level of shear strain.

4 SEISMIC DEMAND AND SOIL STRUCTURE INTERACTION

The previously analysed inertial interaction is of significance not only in cases where the dynamic excitation is applied directly to the structure, but also in seismic design, although the effect of kinematic interaction can be equally important (in some cases even more for deeply embedded foundations).

To estimate the effect of the inertial interaction, let assume the structure to be represented by a single dregree of freedom model of a column with a mass M lumped at a height h over the base and of stiffness K, being

$$M = \frac{\left(\sum m_i x_i \right)^2}{\sum m_i x_i^2} \tag{29}$$

$$K = \frac{4\pi^2 M}{T_0^2} \tag{30}$$

$$h = \frac{\sum m_i x_i h_i}{\sum m_i x_i}. \tag{31}$$

This model can be representative of the behaviour in the first fundamental mode of vibration and accordingly: T_o in equation (30) represents the foundamental period of the structure on a rigid base and x_i is the modal displacement at the height h_i, where we concentrate the mass m_i.

If the soil-foundation system is represented throughout a horizontal stiffness K_x and a rotational stiffness K_α, supposed u_x to be the absolute displacement of the mass, u_b the displacement of the foundation, u_G the motion of the foundation without the structure (as it results from a kinematic interaction only) and y the deflection of the structure, the equations of the motion are the following

$$M\ddot{u}_x + Ky = 0 \tag{32}$$

$$Ky = K_x \left(u_b - u_G \right) \tag{33}$$

$$Khy = K_\alpha \alpha \tag{34}$$

$$u_x = u_b + y + h\alpha. \tag{35}$$

By first substituting (33) and (34) into (35), to eliminate α and u_b, and then (35) into (32), we get the following equation,

$$M \left(1 + \frac{K}{K_x} + \frac{Kh^2}{K_\alpha} \right) \ddot{y} + Ky = -M\ddot{u}_G. \tag{36}$$

that proves that the fundamental period of the whole structure-soil system is given by

$$T = T_o \sqrt{1 + \frac{K}{K_x} + \frac{Kh^2}{K_\alpha}}. \tag{37}$$

As expected, the period increases if the deformability of the soil is taken into account, and for slender structures on weak soils the most important contribution comes from the rotational term in equation (37). This effect will be beneficial or not, depending on the shape of the response spectrum and the initial period T_o. If this period is in the descending branch of the spectrum (as its is usually the case of historic masonry slender towers of our interest) the effect will be beneficial. And further beneficial effects derive from changes in the effective damping (see Gazetas, 1991; Roesset, 1980), resulting in a decrease in forces and accelerations.

REFERENCES

Binda L., Gatti G., Mangano G., Poggi C., Sacchi Mandriani G. (1992). The collapse of the Civic Tower of Pavia: a survey of the materials and structure. Masonry Int., 20, 6, 11–20.

Burland J.B., Jamiolkowski M., Viggiani C. (2003). The stabilization of the leaning Tower of Pisa. Soils and Foundations, 43, 5, 63–80.

Butterfield R. (1980). A simple analysis of the load capacity of rigid footings on granular materials. Journeé de Géotechnique, 128–134.

Cheney J.A., Abghari A., Kutter B.L. (1991). Leaning instability of tall structures. ASCE, JGE, CXVII, 2, 297–318.

Como M. (1993). Plastic and visco-plastic stability of leaning towers. Int. Conf. Physic-Mathematic and Structural Engineering, in memory of G. Krall, Elba.

Como M. (2010). Statica delle costruzioni storiche in muratura. Aracne, Roma, 902 pp.

Desideri A., Russo G., Viggiani C. (1997). La stabilità di torri su terreno deformabile. RIG, 1, 5.

Desideri A., Viggiani C. (1994). Some remarks on the stability of towers. Symp. on Development in Geot. Eng. (from Harvard to New Delhi, 1936–1994), Bangkok.

di Prisco C., Nova R., Sibilia A. (2002). Analysis of soil-structure interaction of towers under cyclic loading. Proc. NUMOG 8, Roma, Pande & Pietruszczak eds., Balkema, 637–642.

Di Tommaso A., Lancellotta R., Focacci F., Romano F. (2010). Uno studio sulla stabilità della Torre Ghirlandina. in La Torre Ghirlandina. Storia e Restauro, Ed. Sossella, Roma, 204–217.

Di Tommaso A., Lancellotta R., Sabia D., Costanzo D., Focacci F., Romano F. (2013). Dynamic identification and seismic behaviour of Ghirlandina Tower in Modena (Italy). 2nd Int. Symposium on Geotechnical Engineering for the Preservation of Monuments and Historic Sites, Napoli.

Federico F., Ferlisi S. (1999). Time evolution of stability of leaning towers. VI Int. Conf. STREMAH, Dresden, Brebbia and Jager Edrs, 485–494, WIT Press.

Federico F., Ferlisi S., Jappelli R. (2001). Safety evolution of masonry leaning towers on deformable soils. Proc. ICSMGE, Istanbul, Balkema, 1, 679–682.

Gazetas G. (1991). Foundation vibrations. In H.F. Fang, Foundation Engineering Handbook, Van Nostrand Reinhold, New York, 553–593.

Hambly E.C. (1985). Soil buckling and the leaning instability of tall structures. The Structural Engineer, 63, 3, 77–85.

Heyman J. (1992). Leaning towers. Meccanica, 27, 153–159.

Lancellotta R. (1993). The stability of a rigid column with non linear restraint. Géotechnique, 33, 2, 331–332.

Lancellotta R. (2009). Aspetti geotecnici nella salvaguardia della torre Ghirlandina. In La Torre Ghirlandina. Un progetto per la conservazione. Ed. Sossella, Roma, 178–193.

Lancellotta R. (2009.a). Geotechnical Engineering. Taylor and Francis, London, 499 pp.

Lancellotta R. (2013). La torre Ghirlandina: una storia di interazione struttura-terreno. XI Croce Lecture, to appear on Rivista Italiana di Geotecnica.

Lancellotta R., Sabia D. (2013). On the role of monitoring and identification analyses on the preservation of historic towers. Keynote Lecture, 2nd Int. Symposium on Geotechnical Engineering for the Preservation of Monuments and Historic Sites, Napoli.

Macchi G. (1993). Monitoring medieval structures in Pavia. Structural Engineering International, 1, 9–9.

Marchi M., Butterfield R., Gottardi G., Lancellotta R. (2011). Stability and strength analysis of leaning towers. Géotechnique, 61, 12, 1069–1079.

Meyerhof G.G. (1953). Bearing capacity of foundations under eccentric and inclined load. Proc. 3rd ICSMFE, Rotterdam, 1, 440–445.

Nova R. (1982). A viscoplastic constitutive model for normally consolidated clay. Proc. IUTAM Symp. on Deformation and Failure of Granular Materials, P.A. Vermeer Luger eds, Delft, 287–295.

Nova R., Montrasio L. (1995). Un'analisi di stabilità del campanile di Pisa. Rivista Italiana di Geotecnica, 2, 83–93.

Paolucci R., di Prisco C., Figini R., Petrini L., Vecchiotti M. (2009). Interazione dinamica non lineare terreno-struttura nell'ambito della progettazione sismica agli spostamenti. Progettazione Sismica, 2, 83–103.

Pepe M. (1995). La Torre pendente di Pisa. Analisi teorico-sperimentale della stabilità dell'equilibrio. Tesi di Dottorato, Politecnico di Torino.

Prager W. (1955). The theory of plasticity: a survey of recent achievements. Proc. Inst. Mech. Eng., 169, 41–57.

Roesset J.M. (1980). Stiffness and damping coefficients of foundations. Proc. ASCE National Convention on Dynamic response of pile foundations: analytical aspects. ASCE, 1–30.

Sabia D. (2012). Rilievo e analisi delle vibrazioni della Torre Ghirlandina di Modena. Contratto di Ricerca tra Comune di Modena e Politecnico di Torino.

Geotechnics and Heritage – Bilotta, Flora, Lirer & Viggiani (eds)
© 2013 Taylor & Francis Group, London, ISBN 978-1-138-00054-4

The Leaning Tower of Pisa

J.B. Burland
Imperial College of London, London, UK

M. Jamiolkowski
Technical Univesity of Torino, Torino, Italy

N. Squeglia
University of Pisa, Pisa, Italy

C. Viggiani
University of Napoli Federico II, Napoli, Italy

ABSTRACT: The Leaning Tower of Pisa is one of the world's best known and most treasured monuments. It was erected in the Middle Ages, at the time of maximum power of Pisa. The Tower is founded on highly compressible soils and started leaning from commencement of its construction. In the 1990s the overhang had reached the value of 4.7 m and was increasing at a rate of 1.5 mm per year; an analysis of the situation showed that a collapse was to be expected within some decades.

After having described the monument and its subsoil, the paper reviews its history and all the data available on the progress of its inclination. Studies and investigations carried out since the early XX century are recalled, and finally the activity of the International Committee appointed in 1990 is reported. After the underexcavation carried out by the Committee, and the decrease of the overhang by about 0,4 m, at present the tower is practically motionless. In conclusion, the possible future scenarios are briefly addressed.

1   INTRODUCTION

The monuments of the Piazza dei Miracoli in Pisa (consisting of the Tower, the Cathedral, the Baptistery and the Monumental Camposanto) were erected in the Middle Ages, in the period of maximum splendour and power of the Pisa Republic. Piazza dei Miracoli is the wonderful symbol of the profound unity prevailing at those times among the religious, spiritual and political powers. The history of Art and the civil history intertwine in the monuments enhancing it, giving them an outstanding character of sign and symbol of the city.

The Leaning Tower (Fig. 1) is one of the best known and most treasured monuments of the world.

Construction is in the form of a hollow cylinder surrounded by six loggias with columns and vaults merging from the base cylinder and surmounted by a belfry. The structure is thus subdivided into eight segments, called "orders".

The external surfaces are faced with masonry of cut stones, the outer one of S. Giuliano marble while the inner one is of various materials. The annulus between the facings is filled with rubble and mortar within which extensive voids have been found. A spiral staircase winds up within the annulus till the 6th order, while two shorter winding staircases lead to the floor and top of the belfry.

The staircase forms a large opening on the south side just above the level of the first cornice, where the cross section of the masonry suddenly decreases. The high stress within this region was a major cause of concern since it could give rise to an abrupt brittle failure of the masonry.

The Tower is founded on weak, highly compressible soils and records indicate that it started leaning since its construction. The movement went on over the centuries and at the end of the first millennium the overhang had reached the worrying value of 4.7 m and was increasing at a rate of 1.5 mm per year.

2 THE SUBSOIL

The ground profile underlying the tower is shown in Fig. 2. It consists of three distinct horizons. Horizon A is about 10 m thick and primarily consists of estuarine deposits, laid down under tidal conditions; as a consequence, a rather erratic succession of sandy and clayey silt layers are found. At the bottom of horizon A there is a 2 m thick medium dense fine sand layer, the so called upper sand.

Horizon B consists primarily of marine clay which extends to a depth of about 40 m. It is subdivided into four distinct layers. The upper layer is a soft sensitive clay, locally known as the Pancone. It is underlain by an intermediate layer of stiffer clay, which in turn overlies a sand layer (the intermediate sand). The bottom layer of horizon B is a normally consolidated clay known as the lower clay. Horizon B is very uniform laterally in the vicinity of the tower.

Horizon C is a dense sand (the lower sand) which extends to considerable depth.

From the geological viewpoint, the lower sands are marine sediments deposited during the Flandrian transgression. The horizon B is formed by Quaternary deposits of marine origin, dominantly clayey, formed at the time of rapid eustatic rise. During the last 10,000 years or so the rate of eustatic rise decreased and the sediments became increasingly estuarine in character. The more recent sediments of horizon A mainly comprise sandy and clayey silt; typically of estuarine deposits, there are significant variations over short horizontal distances. Based on sample descriptions and piezocone tests, the materials to the south of the tower appear to be more silty and clayey than to the north, and the upper sand layer is locally thinner. This is believed to be instigator of the southward inclination of the Tower.

The water table in horizon A is found at a depth between 1 m and 2 m below the ground surface; the latter has an average elevation of 3 m above mean sea level. Pumping from the lower sand has resulted

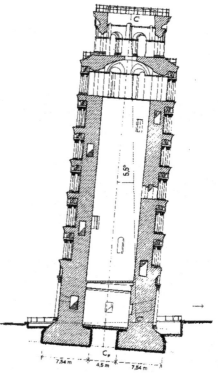

Figure 1.   The Leaning Tower of Pisa.

208

in downward seepage from horizon A with a pore pressure distribution with depth which is slightly below hydrostatic (fig. 3).

The many borings beneath and around the tower show that the surface of the Pancone clay is dished beneath the tower, from which it can be deduced that the average settlement of the monument is not less than 3 m.

Figs. 4 to 8 reports some data on the physical and mechanical properties of the soils, as determined by site and laboratory investigations carried out from 1965 to 1993.

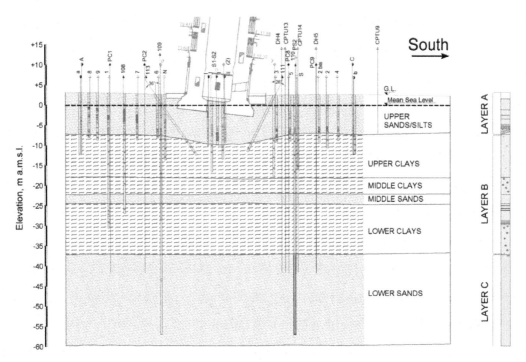

Figure 2.   The subsoil of the Tower.

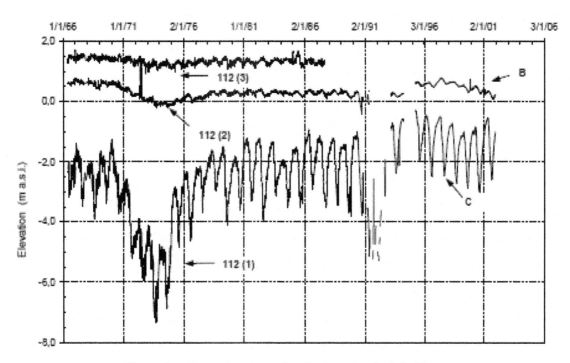

Figure 3.   Ground water regime in the subsoil of the Tower.

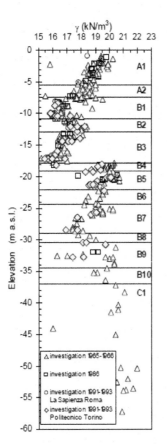

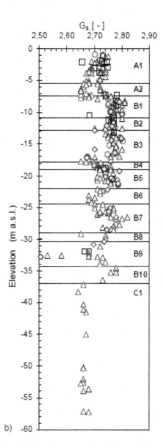

Figure 4.   Unit weight.

Figure 5.   Specific gravity of solid particles.

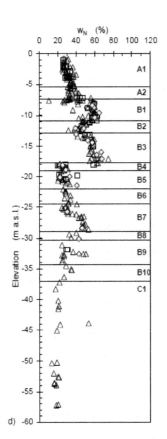

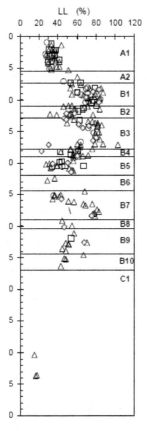

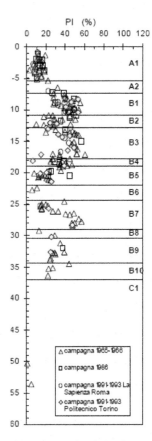

Figure 6.   Natural water content.

Figure 7.   Liquid limit.

Figure 8.   Plasticity index.

3 HISTORY OF THE CONSTRUCTION

Work on the tower began in 1173 (Fig. 9). Construction had progressed to about one third of the way up the 4th order by 1178, when the work was interrupted. The reason for stoppage is not known, but had it continued much further, the foundations would have experienced an undrained bearing capacity failure. As a matter of fact, in 1178 the weight of the tower was around 90 MN, with an average pressure of around 315 KPa on the foundation; under undrained conditions, the bearing capacity of the foundation was of the same order (with an average $c_u = 5.5$ KPa, $q_{lim} \approx 6c_u = 330$ KPa).

Work recommenced in 1272, after a pause of nearly 100 years, by which time the strength of the ground had increased due to consolidation under the weight of the structure. By about 1278, construction had reached the 7th cornice when work again stopped. Once again, had the work continued, the tower would have fallen over. In about 1360 work on the bell chamber was commenced and was completed in about 1370, two centuries after the start of the work. It is known that the tower must have been tilting to the south when work on the bell chamber began, as it is noticeably more vertical than the remainder of the tower. Indeed on the north side there are four steps from the seventh cornice up to the floor of the bell chamber, while on the south side there are six steps (Fig. 10).

Another important detail of the history of the tower is that in 1838 a walkway was excavated around the foundation. This is known as the *catino*, and its purpose was to expose the column plinths and

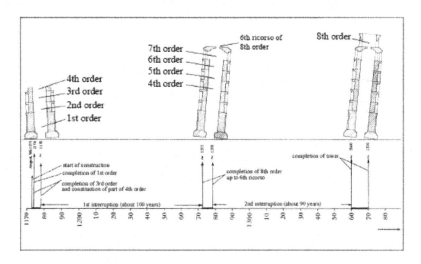

Figure 9. History of the construction of the Tower.

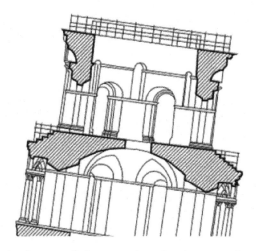

Figure 10. The last correction: At the top of the seventh order there are six steps at south and only four steps at north.

211

foundation steps for all to see as it was originally intended. The operation resulted in an inflow of water, the bottom of the catino being well below the ground water table. As a consequence, since 1838 the catino had been kept dry by continuous pumping.

In the belief that pumping could be dangerous for the stability of the tower, in the years 1934–35 the tower foundation and the soil surrounding the catino were made watertight by injecting cement grout into the foundation masonry and chemical grout into the soil. The intervention succeeded in effectively stopping the water inflow, and since then pumping was consequently interrupted.

4 HISTORY OF THE INCLINATION

A reliable clue on the history of the tilt lies in the adjustments made to the masonry layers during construction and in the resulting shape of the axis of the Tower. Based on this shape and a hypothesis on the manner in which the masons corrected for the progressive lean of the tower, the history of inclination of the foundation of the tower reported in Figure 11 may be deduced. During the first phase of construction to just above the third cornice (1173 to 1178), the tower inclined slightly to the north. The construction stopped for almost a century, and when it recommenced in about 1272 the tower began to move towards the south. When the construction reached the seventh cornice in about 1278, the inclination was about 0.6° towards the south. During the next 90 years the construction was again interrupted and the inclination increased to about 1,6°. After the completion of the bell chamber in about 1370, the inclination went on increasing. Some information on its trend may be obtained by pictures or documents; among them, the value of the inclination deduced by a fresco painted in 1385 by Antonio Veneziano in the Camposanto and the value reported by Giorgio Vasari in 1550. In 1817, when Cresy and Taylor made the first recorded measurement with a plumb line, the inclination of the tower was about 4,9°. In 1859 Rohault de Fleury carried out another measurement, finding a value of the inclination significantly higher than that of Cresy and Taylor. In fact, between the two measurements the walkway surrounding the base of the tower (the so called *catino*) had been excavated to uncover the base of the monument which had sunk into the soil due to a settlement as high as 3.5 m. Digging the catino seriously threatened the stability of the tower, and caused an increase of inclination of approximately 0.5°; furthermore, the rate of inclination increased and the motion changed from retarded to accelerated.

Since 1911 the inclination of the tower has been monitored by different means. It increases slightly more than the rotation of the foundation (fig. 12), implying a steady deformation of the tower body.

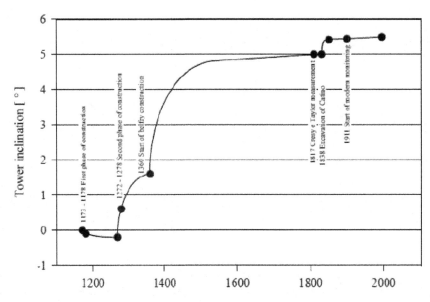

Figure 11.   History of the inclination of the Tower. From 1273 to 1370 the inclination is deduced from the shape of the axis; from 1370 to 1817 from documents and pictures; since 1817, from direct measurements.

212

The long term steady trend is punctuated by two major perturbations: one in 1935 and another one in the early 1970's. The first one was caused by the above mentioned cement grouting into the foundation body and the soil surrounding the catino, carried out to prevent the inflow of water. The second perturbation is related to the pumping of water from deep aquifers, inducing subsidence all over the Pisa plain. The closure of a number of wells in the vicinity of the tower stopped the increase of the rate of tilt.

In any case, even correcting for the anomalous increments occurred in 1935, 1970–73 and some further minor perturbation (fig. 13), it appears that the rate of tilt was steadily increasing and had nearly doubled from 1938 to 1993. In the early 1990s' the inclination was about 5.5°.

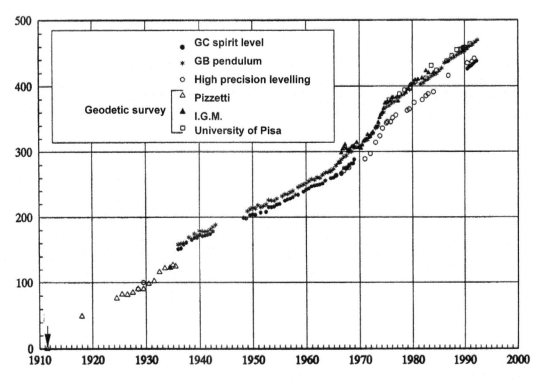

Figure 12.   Increase of the inclination since 1911. GB inclinometer and geodetic measurements include the deformation of the Tower body; GC level and optical levelling are representative of the rotation of the foundation.

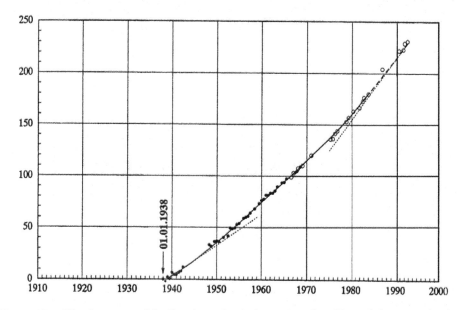

Figure 13.   The increase of inclination after subtracting the effect of the perturbations.

5 PREVIOUS STUDIES AND INVESTIGATIONS

5.1 *From 1902 to 1973*

The first Commission on the tower of Pisa appointed by the Italian Government was a consequence of the worries induced in the public opinion by the collapse of the S. Marco bell tower in Venice, 1902. The Commission carried out a number of investigations, and presented the results in a broad and valuable report, issued in 1912.

A second Commission was appointed in the same year with the task of studying the possible means of stabilising the tower, but did not conclude its work because of the first World War. A new Commission with the same task was nominated in 1924; it included a number of experienced engineers, and developed a solution consisting in widening the foundation of the tower by filling the catino with concrete. The proposal met with strong opposition in Pisa, and another alternative Commission was appointed by the local authorities, to develop different and less intrusive solutions.

In 1927 the Government succeeded in unifying the two Commissions in a new one, which came to the conclusion that the most urgent need was that of sealing the catino. As mentioned before, since 1838 the walkway was kept dry by continuous pumping.

In the years 1934–35 the tower foundation and the soil surrounding the catino were made watertight by injecting 100 t of cement grout into the foundation masonry and 21 m³ of chemical grout into the soil. As already reported, the intervention stopped the water inflow into the catino; the price for this success, however, was a new sudden and marked increase of the inclination of the tower. About 100 years after the excavation of the catino, another intervention carried out with wishful thinking to stabilise the tower again strongly threatened it; this confirms that the way to hell is paved with good intentions!

After the second World War, it became clear that the tower was still moving, in spite of the works carried out in 1935. A permanent Commission was thus appointed in 1949, and among other tasks it had to examine and evaluate a number of design schemes. Though proposed by renowned engineers, all of them were intrusive and not respectful of the historical and material integrity of the monument; with hindsight, it appears very lucky that the Commission did not recommend any of these solutions. Though being "permanent", the Commission was dismantled in 1957; only the inclination measurements were continued. Further solutions were proposed in the following years (fig. 14); it is noteworthy that one of the most intrusive was suggested by the architect N. Benporad, Superintendent of Monuments of Pisa! None of these proposals were considered further.

In 1964 a new and very important Commission was appointed, with the task of preparing the documents of an international competition for the design and implementation of stabilising works. The Polvani Commission, as it is named because of its chairman, included for the first time a group of geotechnical engineers: C. Cestelli Guidi, A. Croce, E. Schultze, A.W. Skempton. To make available a com-

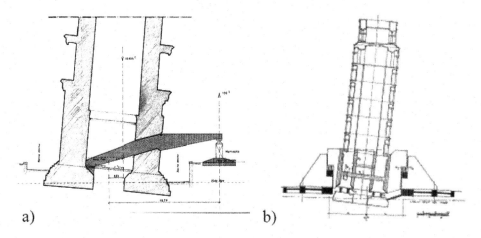

a)   b)

Figure 14.   a) Proposal by Colonnetti (1963); b) Proposal by Benporad and Vannucci (1963).

plete documentation, the Polvani Commission carried out a number of investigations and collected an impressive amount of knowledge.

5.2 *The tender of 1973*

The call for tender was issued in 1973. There was the participation of 22 groups, and 11 among them were admitted to the competition. Prof. Polvani died in 1970, but the Commission, with minor modifications, was charged with judging the competition. The proposals by: Fondedile, Fondisa (fig. 15), Geosonda, Impresit-Gambogi-Rodio (fig. 16) and Konoike (the latter was based on jet-groting the foundation soil) were judged worthy of mention, but no contract was awarded. At that time, it was discovered that the Piazza dei Miracoli was affected by a subsidence process, induced by water pumping from deep wells. This factor was not properly considered in the tender, and this was one of the reasons why the contract was not awarded. Three of the groups which had been mentioned joined together in a Consorzio, and developed a common solution working in connection with the Commission, but eventually nothing was done.

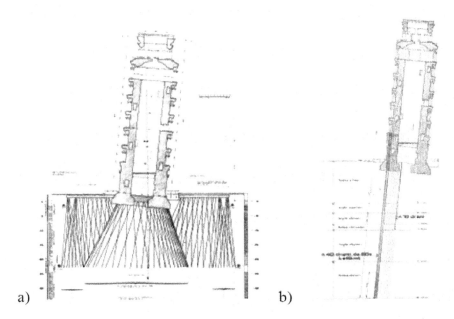

a)   b)

Figure 15.   International tender, 1973. The proposals by Fondedile a) and Fondisa b).

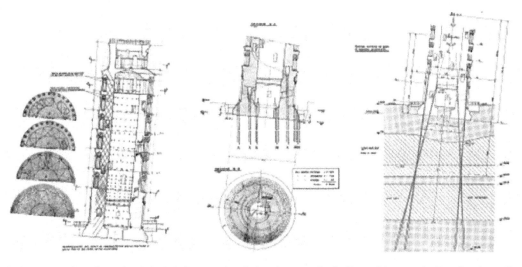

Figure 16.   International tender, 1973. The proposals by Geosonda (left) and Impresit Gambogi Rodio (right).

215

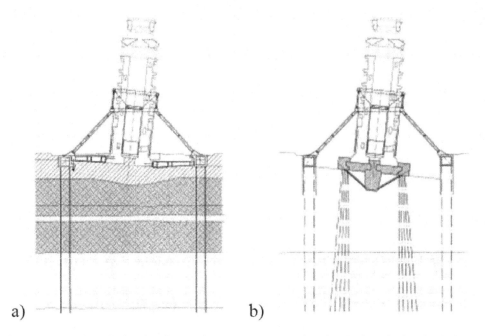

Figure 17.   The solution by the Design Group, 1983. a) *soft* solution; b) *hard* solution.

5.3   *From 1975 to 1990*

In 1983 a Design Group was commissioned by the Ministry of Public Works to design the stabilisation work; they produced a very sophisticated but still rather intrusive solution (fig. 17), that was not definitively approved by the Council of Public Works. In 1988 a technical Committee, entrusted by the Government to study the problem, focused attention on the risk of a brittle failure of the heavily stressed masonry, in addition to the risk of a foundation failure. A failure of the masonry would be sudden, without forewarnings, and therefore potentially very dangerous.

In 1989 another spectacular tower collapse occurred in Italy: that of the Civic Tower of Pavia, with five casualties. As a result, the attention to the safety of the Tower of Pisa increased, and the Government prohibited the access of visitors, following a recommendation of the technical Committee.

The closure of the Tower resulted in a strong public pressure for a rapid reopening, but the restoration experts warned against hasty and insufficiently considered solutions. The Italian Government decided to appoint a further Commission, a truly interdisciplinary International Committee chaired by a geotechnical engineer and formed by art historians, restorers, structural engineers and geotechnical engineers, among whom are the authors of this paper. It had the task of conceiving, designing and implementing the necessary stabilisation works.

6   INVESTIGATIONS AND INTERVENTIONS BY THE INTERNATIONAL COMMITTEE (1990–2001)

6.1   *Analysis of the statics of the tower-subsoil system*

A careful study of the behaviour of the tower led to the conclusion that it was affected by a phenomenon of instability of the equilibrium, known as leaning instability, depending on the stiffness and not on the strength of the foundation soil. To demonstrate leaning instability, the simple conceptual model of an inverted pendulum may be used. It is a rigid vertical pole (Fig. 18) with a concentrated mass at the top and hinged at the base to a constraint that reacts to a rotation with a stabilising moment M_s proportional to the rotation. On the other hand, the rotation induces an offset of the mass and hence an overturning moment M. In the vertical position the system is in equilibrium. Let us now perturb the equilibrium with a small rotation of the pole. If the stabilising moment is larger than the overturning one, the equilibrium is stable; the system returns to the vertical configuration. If the contrary occurs, the equilibrium is unstable; the system collapses. If the two moments are equal, the equilibrium is neutral; the system stays in the displaced configuration. The stability of the equilibrium may be characterised by the ratio $FS = M_s/M$ between the stabilising and the overturning moment.

Modelling the tower as an inverted pendulum, the restraint exerted by the foundation may be evaluated by representing the foundation as a circular plate of diameter D resting on an elastic half space of constants E, v. Defining W and $M = We$ the vertical load and the overturning moment respectively and ρ, α the settlement and the rotation of the foundation respectively (Fig. 19), it may be shown that:

$$\begin{Bmatrix} \rho \\ \alpha \end{Bmatrix} = \begin{vmatrix} \dfrac{1}{k_\rho} & 0 \\ 0 & \dfrac{1}{k_\alpha} \end{vmatrix} \begin{Bmatrix} W \\ M \end{Bmatrix}$$

with:

$$k_\rho = \frac{ED}{i - v^2}; \qquad k_\rho = \frac{ED^3}{6(1 - v^2)}$$

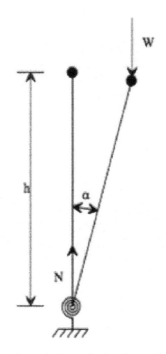

Figure 18.   The inverted pendulum: A simple model of leaning instability.

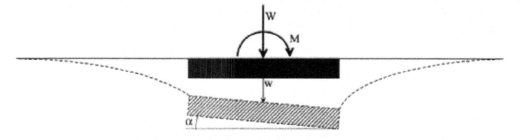

Figure 19.   Eccentrically loaded rigid circular plate resting on an elastic half space.

217

In this simple linear model there is no coupling between settlement and rotation, and the stability of the equilibrium is an intrinsic property of the ground—monument system. It may be characterized by the ratio *FS* between the stabilizing and the overturning moment:

$$FS = \frac{k_\alpha \alpha}{Wh\sin\alpha} = \frac{ED^3}{6(1-v^2)} \frac{1}{Wh}$$

In the case of the tower of Pisa an evaluation of *FS* may be obtained by the knowledge of the settlement of the tower, $\rho \geq 3$ m. Being $k_\rho = W/\rho$, one gets $E/(1-v^2) \leq 2.85$ MN/m^2. Accordingly, with h = 22.6 m (height of the centre of gravity of the tower) and W = 141.8 MN (weight of the tower), *FS* ≤ 1.12. Even this simplistic linearly elastic subsoil model allows the important conclusion that the tower is very near to a state of neutral equilibrium. The continuing movement, made possible by the state of neutral equilibrium, is controlled by ratchetting following cyclic actions such as the fluctuations of water table in Horizon A. Of course, creep has also some influence on the process.

The relationship between the stabilising moment $M_s = k_\alpha \alpha$ and the rotation α may be linearized over a short interval, but it is certainly non linear and approaches asymptotically a limiting value of M_s. In a case such as that of the Leaning Tower, that is on the verge of instability, consideration of non linearity appears mandatory. As a matter of fact, centrifuge experiments by Cheney *et al.* (1991) (fig. 20) show coupling between settlement and rotation, non linearity and strain hardening plasticity.

The leaning instability of the Tower has been investigated by a number of different approaches, including small scale physical tests at natural gravity and in the centrifuge, and Finite Element analyses based on different constitutive models of the subsoil. The analyses led to the conclusion that the gradual increase of the inclination would have ended in a collapse. Another very significant conclusion was that a decrease of the inclination, even a relatively minor one, results in a substantial increase in the safety against leaning instability.

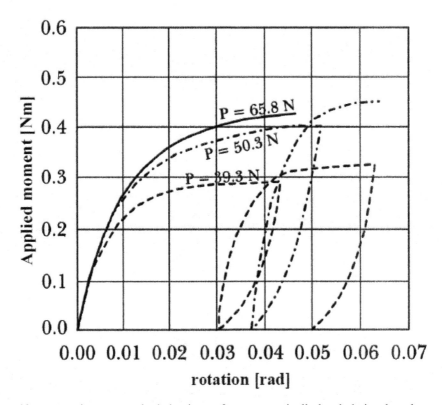

Figure 20. Centrifuge experiments on the behaviour of an eccentrically loaded circular plate on clayey subsoil (Cheney *et al.*, 1991).

To clarify this conclusion, reference may be made again to the inverted pendulum model. In a non linear mode, the relation between loads and displacements has to be expressed in incremental form:

$$\begin{Bmatrix} \partial\rho \\ \partial\alpha \end{Bmatrix} = \begin{vmatrix} \dfrac{1}{k_\rho} & \dfrac{1}{k_{\rho\alpha}} \\ \dfrac{1}{k_{\alpha\rho}} & \dfrac{1}{k_\alpha} \end{vmatrix} \begin{Bmatrix} \partial W \\ \partial M \end{Bmatrix}$$

The increments of displacement depend on the load increments, the current state of load and the load history. Hence the factor of safety depends on the current state of stress and stress history.

It may be seen in fig. 20 that a decrease of the inclination, which involves the unloading branch of the curve, strongly increases the stiffness of the ground—foundation system, and hence the stability. This generated the idea that a decrease of inclination could be used to stabilize the tower.

6.2 *Temporary interventions: Lead counterweights*

Fully aware that a long time was needed to conceive, design and implement the permanent stabilization measures, the Committee took an early decision to implement temporary and fully reversible interventions to slightly improve the safety against overturning, and to gain the time to properly devise, design and implement the permanent solution. A total of 6.9 MN of lead ingots were installed between May 1993 and February 1994 on the north edge of the base of the Tower (fig. 21). They induced a change of inclination of 33″ by February 1994; by the end of July it had increased to 48″ and eventually to 53″. By February 1994 the average additional settlement of the tower relative to the surrounding ground was about 2.5 mm. The settlement and rotation produced by the counterweight had been predicted by the finite element model; the agreement between prediction and observation was satisfactory, increasing the confidence in the model. An event of the utmost importance is that the progressive southward inclination of the Tower came to a standstill.

These observations allow an important experimental confirmation of the intended stabilization approach. The tower reacted to the application of the lead weights with a rotational stiffness:

$$k_\alpha = \frac{43.5\,MNm}{52''} = 172,548\,MNm$$

The factor of safety therefore increased to:

$$FS = \frac{k_\alpha}{Wh} = \frac{172,548}{141.8x22.6} = 54$$

Figure 21. The provisional intervention by lead ingots counterweight.

As a matter of fact, after the application of the counterweight the tower remained essentially motionless for over three years, apart from the seasonal cyclic movements.

6.3 *Temporary interventions: Ground anchors*

The administrative life of the International Committee was somewhat difficult; for various reasons its continuation was repeatedly in doubt and its activity was repeatedly interrupted for periods up to many months. In these conditions there was a widespread fear that the Committee could dissolve, as all the preceding Commission had done, but this time leaving the stack of lead ingots on the tower for a period difficult to foresee, but certainly of some years or even decades. Essentially for this reason, a medium term temporary scheme was developed to replace the lead weights with ten tensioned steel cables anchored in the lower sands at a depth of over 40 m (fig. 22). Apart from the advantage of being invisible, an additional benefits of this scheme was the increased lever arm that would give a stabilizing moment larger than the lead ingots. The major problem of the ten anchors solution was that the anchors had to be connected to the Tower foundation through a ring beam to be constructed below the floor of the catino, and this involved an excavation around the tower below ground water level, an operation of the utmost delicacy. After a careful comparison of different possibilities, it was decided to employ local ground freezing immediately below the catino floor but well above the Tower foundation level. Following investigations by drill cores, it was discovered that below the catino floor there is a concrete bed of 1 m thickness, set in place partly in 1837 and partly in 1935. Some cracks at the interface between the concrete and the tower foundation, led to the conclusion that the two bodies were not connected; as a consequence, the volume variations of the frozen soil during freezing and thawing were expected not to influence the tower.

Without going into the details of the operation, it can be reported that freezing was commenced on the North side and the northern sections of the ring beam were successfully installed. They were connected to the foundation by means of stainless steel rods, cemented in the foundation masonry.

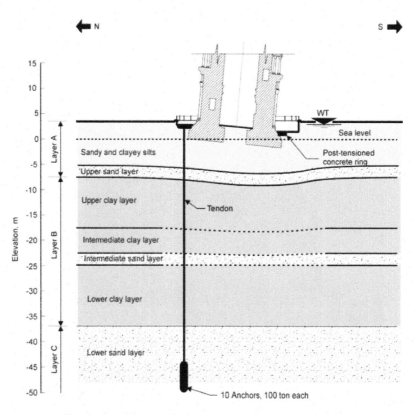

Figure 22.   The medium term solution of ten anchors.

220

During these operations, the water tightness of the catino was partly destroyed and two pumps had to be installed to prevent the flooding; since a sand layer was provided below the ring beam sections, the system worked as a ground water level control.

During the excavation, a number of short steel tubes of the diameter of 50 mm emerging from the tower foundation were discovered. They represent the intake of the injection holes executed in 1934–35 to make watertight the foundation, and left in place. In spite of their short length (about 0.5 m, half cemented in the masonry and half protruding), they represent a connection between the catino and the tower.

In September 1995 freezing was commenced on the south-west and south east sides, and the Tower began to rotate southward; the movement was also affected by an attempt of installing some micropiles at the south boundary of the catino. After some attempts of controlling the rotation by the application of further lead weights at north, the operation was abandoned.

6.4 *Final intervention*

The Committee had developed a deep insight into the behaviour of the tower, through the interpretation of its history, the scrutiny of the measurements taken in the last century and the analysis of the phenomenon of leaning instability. After a comprehensive discussion, it was concluded that a decrease of the inclination of the Tower by half a degree (1800 arc seconds, *i.e.* around 10% of the inclination in 1990) would be sufficient to stop the progressive increase of inclination and to substantially improve the stability conditions. At the same time, such a reduction was considered small enough not to be perceived at a first glance. The decrease had to be obtained by inducing a differential settlement of the tower opposite to the existing one, by acting on the foundation soil and not on the tower. Among other advantages, such a solution is perfectly respectful of the formal, historic and material integrity of the monument.

The Committee studied in detail three possible means to achieve the decrease of the inclination: (i) the construction of a ground pressing slab to the north of the tower; (ii) the consolidation of the Pancone clay north of the Tower by electro-osmosis, and (iii) the controlled removal of small volumes of soil beneath the north side of the foundation (underexcavation).

All three approaches were the subject of extensive numerical modelling.

A large field experiment of electro-osmosis (Squeglia, Viggiani, 2003) showed that the process cannot be completely controlled, and dangerous phenomena such as pore pressure increase may occur. For this reason the use of electro-osmosis was ruled out.

Small scale model tests of underexcavation at natural gravity and in the centrifuge, in addition to numerical analyses, gave a favourable response, encouraging the Committee to undertake a large scale experiment, to develop the field equipment and explore the operational procedures. For this purpose a 7 m diameter eccentrically loaded instrumented footing was constructed in the Piazza and subjected to underexcavation (fig. 23).

The trial was very successful; it proved that it was possible to steer the footing and act on the rate of rotation by varying the position and the intensity of soil extraction. The numerical modelling and the centrifuge tests had revealed the existence of a critical line beyond which the effect of the underexcavation is detrimental; in fact, during the field trial, an overenthusiastic excavation beyond that line actually produced an increment of the inclination thus confirming the prediction.

The Committee was aware that all the investigations carried out might be not be completely representative of the possible response of a tower affected by leaning instability; therefore it was decided to implement a preliminary and limited ground extraction beneath the Tower itself, to observe its response. To prevent any unexpected adverse movement of the monument, a safeguard structure was necessary. It consisted of two sub-horizontal steel stays (fig. 24), connected to the Tower at the level of the third order and to two anchoring frames located some 100 m apart; it was capable of applying to the Tower a stabilising moment, but only if needed. The safeguard structure was installed in December 1998.

The preliminary underexcavation intervention was carried out between February and June, 1999, operating with 12 inclined drill holes and removing a total of 7 m³ of soil, 71% of which was north

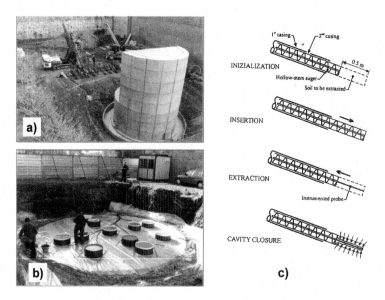

Figure 23. a) The underexcavation trial field. b) Installation of pressure cells on the foundation plane. c) The underexcavation procedure finally developed.

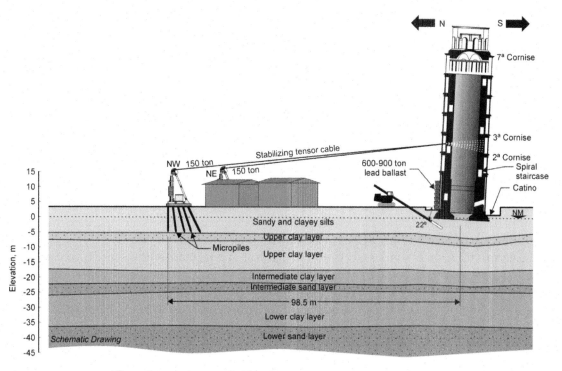

Figure 24. Scheme of the safeguard structure with steel stays.

of the tower and 29% from beneath the foundation. The extraction of soil was very gradual, with the removal of only 22 dm^3 in each single underexcavation. The movements of the tower were monitored by a comprehensive system of geodetic survey and transducers, allowing the acquisition and processing of the data in real time. An observational approach was adopted; the program of soil extraction was decided daily on the basis of the observed movements of the previous day.

The tower rotated northward by 90 seconds of arc till June 1999, when the operation ceased; by mid-September the rotation had increased to 130″. At that time three of the 97 lead ingots were removed, and since then the tower exhibited negligible further movements.

After the very positive results of the preliminary underexcavation, the Committee went on steadily to the full underexcavation (fig. 25). It was carried out between February 21, 2000 and June 6, 2001,

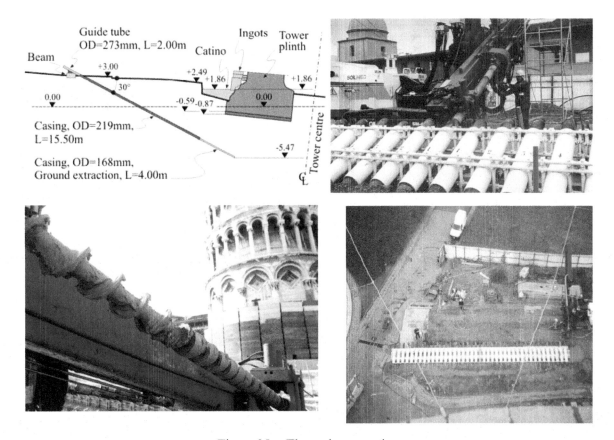

Figure 25.   The underexcavation.

with 41 holes, removing a total of 38 m³ of soil (70% below the catino, *i.e.* outside the perimeter of the foundation) involving 1737 single extractions. In the same period all the lead ingots were progressively removed. In June 2001 the steel cable stays were dismantled, without ever having been operated.

The goal of reducing the inclination of the tower by half a degree has been fully attained (fig. 26). The intervention brought the tower back to the position it had at the beginning of the XIX century, just before the excavation of the catino (fig. 27). It can be seen as a reparation to the incautious undertaking of the architect Gherardesca, and there is a kind of poetic justice in repairing the negative effect of an imprudent excavation with another well-conceived and carefully conducted excavation.

It is important to add that the study of the movements of the Tower revealed that the oscillations of the ground water table consequent to heavy rainfalls exerted a small negative influence on the monument. As a matter of fact, the ground water table at south of the foundation is around 0.4 m higher than that at north, so that the net result of the underpressure on the Tower is a small stabilizing moment. During intense rainfall events the two levels tend to equalize (fig. 28), thus producing a small overturning moment on the monument; it is believed that the cumulative effects by ratchetting of these repeated impulses has been one of the factors producing the steady increase of inclination in the long term.

As a final intervention, a drainage system was thus installed in April 2001 at north of the Tower (fig. 29), essentially aimed at stabilizing the groundwater level in the vicinity of the Tower. It produced a further reduction of the inclination of around 60 seconds of arc, that can be clearly detected in fig. 26. After underexcavation and drainage, at present the Tower shows small cyclic movements connected to environmental actions, and there may be still a very slow movement northwards.

6.5   *Future scenarios*

More than ten years have elapsed since the end of underexcavation and the installation of the drainage system, and as mentioned above the tower is still moving northwards. In spite of this very satisfactory

223

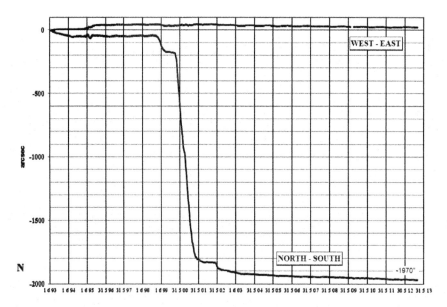

Figure 26.   Rotation of the Tower foundation since December 1998.

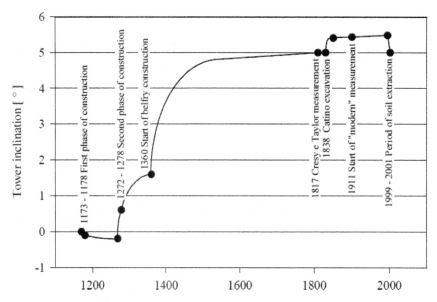

Figure 27.   History of the rotation of the Tower from the time of its construction.

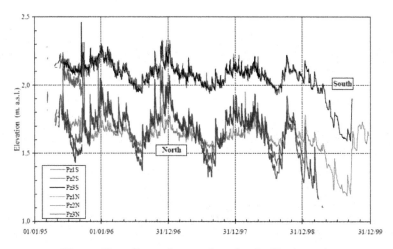

Figure 28.   Groundwater elevation in Horizon A.

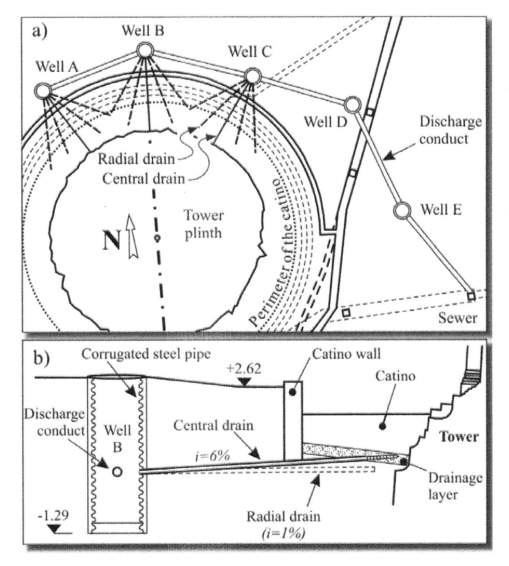

Figure 29.   Drainage system to control the ground water table in Horizon A.

situation, a prediction of the future behaviour is not simple, because of the complexity of the phenomena controlling it. Two possible scenarios may be envisaged: an optimistic and a pessimistic one (fig. 30).

In the optimistic scenario the progressive increase of the inclination, affecting the tower since the construction, has been definitively stopped and the monument keeps motionless, apart from the cyclic movements connected to the environmental action, such as the daily sun irradiation and the seasonal groundwater table fluctuation. If in future the drainage system is kept effective by proper maintenance, the main cyclic action—the fluctuations of the water table—will be strongly attenuated if not eliminated.

The pessimistic scenario sees the tower staying motionless for a period, the honeymoon, of some decades, followed by a resumption of the southward rotation with a steadily increasing rate, and approaching after a long time the value it had experienced at the end of XX century. It is worth pointing out that, should the pessimistic scenario actually occur, the tower would reach the inclination it had before underexcavation in a time of the order of 300 years; this leaves ample margin for further interventions, if needed, possibly repeating the underexcavation.

In order to provide a reference in the debate about future scenarios, a rather sophisticated numerical analysis of the tower-subsoil system is being carried out. The analysis takes into account the three dimensional geometry, the history of construction and all the details of the interventions carried

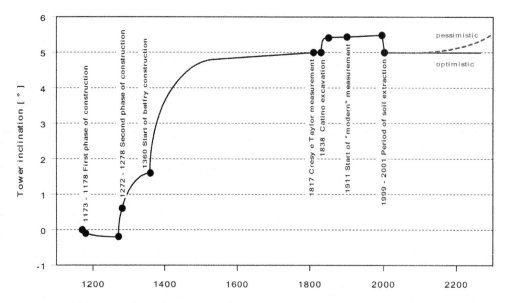

Figure 30.   Possible scenarios of the future behaviour of the Tower.

out since the XIX century; the constitutive model employed includes volumetric (but not deviatoric) creep.

The first results obtained support qualitatively the pessimistic scenario, in the sense that a honeymoon period followed by a resumption of the southward movement is foreseen. In any case, at the present time a comparison with the actual observations of the tower behaviour seems to indicate that the prediction is rather conservative.

7   CONCLUDING REMARKS

The stabilisation of the Tower of Pisa has been a very difficult challenge for geotechnical engineering. The tower is founded on weak, highly compressible soils and its inclination had been increasing inexorably over the years to the point at which it was about to reach leaning instability. Any disturbance to the ground beneath the south side of the foundation is very dangerous. Therefore the use of conventional geotechnical processes at the south side, such as underpinning, grouting, etc., involved unacceptable risk.

The internationally accepted conventions for the conservation and preservation of valuable historic buildings, of which the Pisa Tower is one of the best known and most treasured, require that their essential character should be preserved, with their history, craftsmanship and enigmas. Thus any intrusive interventions on the tower had to be kept to an absolute minimum and permanent stabilisation schemes involving propping or visible support were unacceptable and in any case could have triggered the collapse of the fragile masonry.

The technique of underexcavation provided an ultra soft method of increasing the stability of the tower, which is completely consistent with the requirements of architectural conservation. Different physical and numerical models have been employed to predict the effects of soil removal on the stability. It is interesting to point out that some mechanisms (as, for instance, the occurrence of a critical line beyond which the underexcavation becomes dangerous) are predicted by physical modelling and by the FEM analyses, while these are missed by the simplified Winkler type models.

The preliminary underexcavation intervention, only undertaken once the Commission were satisfied by comprehensive numerical and physical modelling together with a large scale trial, has demonstrated that the Tower responds very positively to soil extraction. The final underexcavation has attained the target of reducing the tilt of the Tower by half a degree.

226

It is believed that the geotechnical stabilisation has been finally attained; monitoring the behaviour of the monuments for the forthcoming few years will confirm it.

REFERENCES

Burland, J.B., Jamiolkowski, M.B., Lancellotta, R., Leonards, G. & Viggiani, C. (1993). The Leaning Tower of Pisa, what is going on? *ISSMFE News* 20(3).

Burland, J.B., Jamiolkowski, M.B. & Viggiani, C. (2000). Underexcavating the Tower of Pisa: Back to the future. In *GEOTECH-YEAR 2000, Developments in Geotechnical Engineering*, Bangkok, Thailand, Balasubramaniam, A.S. et al. Eds, pp. 273–282.

Cestelli Guidi, C., Croce, A., Skempton, A.W., Schultze, E., Calabresi, G. & Viggiani, C. (1971). Caratteristiche geotecniche del sottosuolo della Torre, in *Ricerche e studi sulla Torre pendente di Pisa ed i fenomeni connessi alle condizioni d'ambiente*, IGM, Firenze, I, pp. 179–200. In Italian.

Cheney, J.A., Abghari, A. & Kutter, B.L. (1991). Leaning instability of tall structures. *Journal of Geotechnical Engineering*, ASCE, 117(2): 297–318.

Cresy, E. & Taylor, G.L. (1829). Architecture of the Middle Ages in Italy: illustrated by views, plans, elevations, sections and details of the cathedral, baptistry, leaning Tower of campanile and campo santo at Pisa from drawings and measurements taken in the year 1817, London: published by the Authors.

Croce, A., Burghignoli, A., Calabresi, G., Evangelista, A. & Viggiani, C. (1981). The Tower of Pisa and the surrounding square: recent observations, in *X International Conference on Soil Mechanics and Foundation Engineering*, Stocholm, III, pp. 61–70.

Desideri, A., Russo, G. & Viggiani, C. (1997). Stability of towers on compressible ground, *Rivista Italiana di Geotecnica*, 31(1): pp. 5–29.

Desideri, A. & Viggiani, C. (1994). Some remarks on the stability of towers, in Symposium on development in *Geotechnical Engineering, from Harvard to New Delhi 1936–1994*, Bangkok.

Esrig, M.I. (1968). Pore pressures, consolidation, and electrokinetics, *Journal of Soil Mechanics and Foundation Division*, ASCE, 94(SM4): 899–821.

Johnston, G. & Burland, J.B. (2004). Some historic examples of underexcavation. *Int. Conf. on Advances in Geotechnical Engineering, The Skempton Conference*, London, UK, Jardine, R.J. Potts, D.M. and Higgins, K.G. Eds, Vol. 2, pp. 1068–1079.

Lancellotta, R. & Pepe, M. (1998). On the stability of equilibrium of the Leaning Tower of Pisa, in *Atti Sc. Fis. Accademia delle Scienze*, Torino, 132, pp. 1–11.

Rohault De Fleury, (1859). Le Campanile de Pise, *in Encyclopedie de l'Architecture,* Paris: Bance.

Terracina, F. (1962). Foundation of the leaning tower of Pisa, *Geotechnique*, XII, n. 4, pp. 336–339.

Trevisan, L. (1971). Caratteri geologici, chimici e mineralogici del sottosuolo della Torre e nei pressi di essa, in *Ricerche e studi sulla Torre pendente di Pisa ed i fenomeni connessi alle condizioni d'ambiente*, IGM, Firenze, I, pp. 151–164. In Italian.

Viggiani, C. & Squeglia N. (2003). Electroosmosis to stabilize the leaning tower of Pisa, *Rivista Italiana di Geotecnica* 37(1): 29–37.

Geotechnics and Heritage – Bilotta, Flora, Lirer & Viggiani (eds)
© *2013 Taylor & Francis Group, London, ISBN 978-1-138-00054-4*

Strengthening of the San Marco bell tower foundation in Venice

G. Macchi & S. Macchi
Studio Macchi, Milan, Italy

M. Jamiolkowski & V. Pastore
Studio Geotecnico Italiano, Milan, Italy

D. Vanni
Trevi S.p.A., Cesena, Italy

ABSTRACT: The San Marco bell tower dates back to 12th Century. On July 14th, 1902 the tower bricks masonry collapsed and 14,000 tons of debris dropped on the San Marco square causing the breakdown of the NE corner of the Marciana library and of the Sansovino's loggetta. The reconstruction started in 1903 and was completed in 1908. The designers decided to enlarge the stone masonry foundation from the original area of 222 m^2 to 407 m^2. The new stone masonry was locked together to the old one. Since early '50's, some cracks were discovered in the foundation block, indicating the detachment of the new masonry from the pre-existing one. The progressive widening of the cracks opening as been monitored since their early discovery and recent data triggered the decision of the Ministry of Infrastructures and Trasportation, through the Concessionary Consorzio Venezia Nuova, to reinforce the foundation. The intervention consists of inserting at two levels along the perimeter of the plinth prestressed titanium rebars, to increase the overall flexural stiffness of the foundation and stop further cracks opening. The intervention required excavations below the ground water level in soft/loose lagoon deposits within an extremely important monumental area and in presence of buried archeological remains, many of which were unknown. In such very sensitive environment, the works required severe precautionary measures to prevent even little settlements of the tower and surrounding monuments during excavation, dewatering and retrieval of the buried archeological remains. To achieve this goal a large number of deep in-place mixed columns was foreseen, to create both watertight retaining walls and bottom plugs of the pits to be excavated. The paper will briefly illustrate the main features of the strengthening project and present a detailed description of geotechnical works that made the implementation of the project feasible.

1 RECONSTRUCTION OF HERITAGE BUILDINGS

The reconstruction of a famous heritage building occurs hardly ever, being the overall collapse of a 800 year old monument a very rare event. Moreover, the advanced knowledge of Heritage Conservation is not in favor of a reconstruction, generally seen as an historical fake, far from the aim of conservation heritage monuments as evidence of our civilization.

The historic monuments are the living witnesses of their age and should be regarded as a common heritage. The joint responsibility to safeguard them for future generations is recognized and it is our responsibility to pass them on in the full richness of their authenticity. Thus the restoration of heritage buildings need be an interdisciplinary process more than a simple reconstruction "à l'identique", in order to preserve at the best all the different values (artistic, technical, historical).

2 HISTORICAL BACKGROUND

2.1 *The reconstruction*

However, when in 1902 the San Marco bell tower gradually collapsed during the restoration works, the unanimous wish of the public opinion, Venetians and foreign art lovers all were in favour of a reconstruction *"where and as it was"* mainly to keep the universally known landscape mark view of the Piazza (Figure 1).

In 1902 a multidisciplinary committee was appointed to plan the reconstruction, addressing all the relevant features including the historical, artistic, and technical aspects of the problem. The collapse was due to the poor state of conservation of the tower masonry, built between the 10th and the 11th century; previous damages due to lightning occurred in 1489 and 1745 and were not properly repaired.

Moreover, the construction of the upper marble cell and of the spire did considerably increase the original weight, and a differential settlement of the foundation was leading the monument towards the collapse. The differential settlement was about 10 cm in the N-E direction, with an out-of-plumb of the tower of 80 cm (Donghi, 1913). Such settlement was not regarded as a real threat so that the reconstruction began on the remains of the old foundation block.

The works, however, did not follow the previous design, and the total weight was reasonably decreased. Moreover, the commission was doubtful about the resistance of the foundation hence the size of the base was increased from 222 m^2 to 407 m^2. Such amount of the external increase of the stone base suggests that for the foundation strength the new engineers used far different criteria than the original design.

The work procedure was analogous to the original one: inside a wide wooden pile fence 3086 additional larch piles were driven, close to each other, to consolidate the soil of the additional areas for a depth of 4 to 7.60 m. Once stabilized the piles with the cement, three thick levels of wood planks provided the horizontal level for the new stones of the pyramidal foundation block. The new stones were carefully designed and cut in such a way to firmly interact and form a solid block. The unit load on the soil was reduced from 900 to 400 kPa.

The safety was considerably improved. However, only a few decades later, the foundation began to show a faulty behavior.

Figure 1. The San Marco bell tower.

2.2 Survey of the bell tower in 1993–1999

Quite evident cracks on the trachyte steps at the level of the Piazza called the attention of the technicians. Some cracks showed a clear direction connected with shear stresses; not surprising though, due to the well known trachyte fragility. However, a first alarm arose, and the new cracks were taken under control by the Procuratoria. Six trenches excavated on three sides of the tower showed that several sub-vertical cracks already appeared on the external surface of the steps of the foundation block (Figure 2).

In 1955 a systematic monitoring performed by the University of Padova with 22 mechanical Whittemore extensometers showed an increasing linear movement with time of the cracks, reaching the thickness of 1 mm in 1975.

The damage seemed actually very small and the monitoring was stopped that same year, hoping in a stabilization with time. Any strengthening intervention was delayed, in spite of some concern raised by the Commission. The cracks showed the unambiguous effects of bending and shear, as a result of the insufficient thickness of the added stone platforms. An outer reinforced concrete chain and deep steel connections of the stones were suggested (but not implemented).

As a consequence of the tragic, unexpected collapse of the Civic Tower of Pavia, occurred in 1989, the Authors were called to a general survey of the San Marco bell tower as well as of other monuments in jeopardy.

A complete structural survey of the tower was carried out by ISMES, and an automatic monitoring system was installed on several cracks. The system was able to detect, in real time, also the movements of several critical points of the tower, which were compared with the movements calculated by a Finite Element Model. A system of vertical cracks of the shaft was clearly found at the height of 25 m from the soil. A clear explanation was identified in the temperature cycles of the external surface, causing several cracks of limited depth which although had to be carefully taken into account in the overall study, did not represent a serious danger for the good brickwork of the tower.

Alarming results arose from the vertical state of stress measured in 49 critical points of the tower with flat-jacks and the values were considered significantly higher than those calculated in the lower zone of the tower.

The first comparison was evaluated between the FEM prevision for the vertical stresses in the masonry and the corresponding results obtained by Donghi in his analysis carried out in 1913. The FEM mean stress results, including the increased foundation, were only 4% higher than those calculated by Donghi and at a first sight, the comparisons between the two methods seemed satisfactory. Even the spread among the experimental values obtained by flat jacks for the stress of the shaft at different levels was generally low. The experimental error of flat-jacks was estimated within 0.5 MPa.

Figure 2.   Shear cracks in the stone foundation.

On the contrary it was very much surprising the critical concentration of vertical stresses found by the flat-jacks in the shaft lower part, at the four corners of the section.

Such a dangerous stress concentration at the shaft angles is due to the deformability of the base in comparison to the great stiffness of the box section of the shaft. This result was found by the flat-jacks and confirmed by the FEM analysis.

Figure 3 shows how the base of the tower (the shaft at the height of 0.65 m over the pavement of the tower) tends to take the shape of a cup under the effect of dead load.

These local values are about three times the values of the first analysis, and perhaps more than half the strength of the masonry.

The stress values are correlated to the shape of the base of the tower, and therefore to the systematic sub-vertical cracks found on the steps of the foundation block since 1955. The monitoring of the cracks, suspended in 1975, was suddenly resumed, giving surprising results: instead of a stabilization, all the cracks showed a linear increase with the time, reaching a thickness of about 2 mm, twice the value measured in 1955 (Figure 4). All the data confirmed that the differential settlement of the foundation block was a movement in progress. The reason is the insufficient stiffness of the stone additions done at the time of the reconstruction. The corresponding increase of the cracks and of the compression stress of the bricks is a progressive dangerous phenomenon able to produce at least local collapses of the "new" bell tower.

The consequence is that vertical stresses measured by flat jacks near the base corners are between 2.34 and 3.24 MPa, against the mean value 0.8 MPa found by Donghi in the initial analysis. An effective intervention was urgently needed to stop the process and avoid any unexpected effects with time.

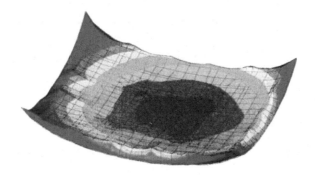

Figure 3.   "Cup shape" reached by the foundation block.

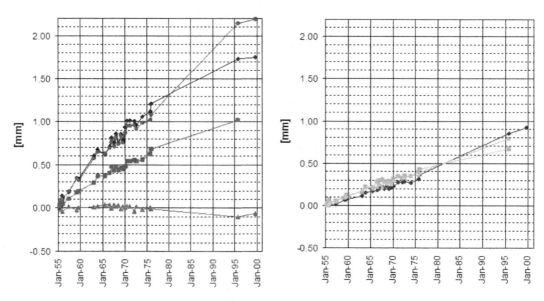

Figure 4.   Time history of the opening of cracks in the stone foundation.

232

Well aware of the time that would be needed to conceive and implement the permanent stabilzation, the Committee required the installation of a very efficient monitoring system and the continuous control of the micro-movements of the tower and of its cracks.

2.3 *The project of the overall intervention*

Further investigations were carried out across the foundation block to ensure that the connection between the ancient block and the stones added in 1903 was efficient and uniform despite any possible cracks.

Six "integral samplings" 50 mm in diameter were drilled, crossing at 45° the surface of the connection between the ancient and the new stones (Figure 5). Analysis on both the cores and on the holes surface proved that the connection between the two parts was perfectly preserved, finally making evident that the reason for the dangerous phenomena discovered at the base of the tower was due to the set of cracks monitored for 45 years on the stone steps. The intervention could therefore be based on very sound data.

An innovative intervention was proposed, gradual, reversible, durable and avoiding any invasive prostheses, as would have been the reinforced concrete ring suggested in 1955. The intervention consisted in leaving unchanged the structure, both in elevation and in foundation, but applying a horizontal precompression with two levels of titanium chains external to the foundation stone block. The system of forces applied to the bars by jacks would be sufficient to stop any further opening of the cracks, as shown by a FEM modeling of the intervention, and should be permanent.

The improvement introduced was the application of "dynamometric bars"; a technique already applied successfully in 2002 to stop the out-of-plane movement of the Façade of the Basilica di San Pietro in Rome towards the Piazza. The principle takes advantage of the friction forces already acting between the stone blocks (and already counteracting the relative movement), only adding a limited force (able to stop the movement of the cracks) with the titanium bars.

Only rather small forces are applied, with limited disturbance to the monument. Their value is able to produce a small reduction of the cracks, and may be adjusted, if necessary, with time under strict monitoring (Figure 6).

In order to provide the maximum level of the bars durability, (considering they have to work in a highly corrosive environment), "pure titanium" ASTM B265 grade3 was used. An alternative use of titanium alloys would have provided higher strength, but less corrosion resistance. This choice gives priority to durability, as the corrosion velocity in marine environment will be technologically zero. The strength of the material was 450 MPa and the ultimate elongation 18%.

Each chain is made of 2 bars of 60 mm of diameter, free inside a protective plastic tube of 450 mm of diameter. Bars are therefore free from contacts with the foundation block and only exert their end forces to the anchor elements at the corners. In this way the block is submitted to 4 diagonal forces

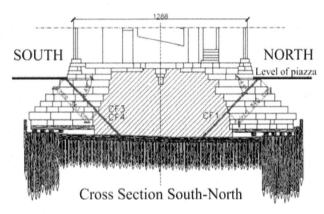

Figure 5.   45° integral samples across new and original stone masonry.

Figure 6.   Installation of dynamometric titanium bars.

Figure 7.   Titanium corner elements.

providing the necessary inward pre-stress. For the upper chain the support elements are made with titanium plates applied to the stones (a detail of corner titanium elements is shown in Figure 7). High strength granite blocks are used for the lower chain level. The bars themselves are used to precompress the granite blocks, in order to avoid micro-cracks under the high forces applied. An appropriate application of the bars required difficult works, in order to provide holes and tubes for the bars and close possible micro-cracks. The access of the workers and of the equipments under the level of the Piazza was guaranteed and made safe by the construction of 4 underwater chambers requiring innovative techniques.

The overall intervention project was conceived and developed by G. Macchi and S. Macchi, while all the geotechnical interventions were defined by M. Jamiolkowski, V. Pastore and late C. Mascardi.

Trevi S.p.A., entrusted from the Main Contractor (Sacaim S.p.A.) for the soil consolidation activities, cooperated with the designers for all the geotechnological aspects.

3   SUBSOIL CONDITIONS AND WATER LEVELS

The intervention took place within an extremely sensitive geotechnical framework, characterized by the presence of soft and fines lagoon sediments, with extensive historic fill, especially in central Venice.

The typical soil profile in the area of San Marco Square was identified on the basis of geotechnical boreholes and CPTU test carried out in four campaigns, carried out in 1993, 1997, 2005 and 2006. The investigations consisted of geotechnical boreholes with recovery of undisturbed and remolded samples.

The identified average soil profile can be summarized as follows:

- From San Marco Square pavement level (+0.90 m on msl) to approximately 5.00 m depth, fill consisting of sandy-clayey silts (locally medium to fine sand with silt). Masonry debris, trachyte blocs, wooden piles unreinforced concrete are present with variable thickness (from few cm to 2.6 m within the area of interest).
- Below fill to 6 ÷ 7 m depth from G.L. a layer of soft sandy clayey silt and/or silty clay with organic debris and peat (as a general indication, point resistance values from CPTU tests are in the order of 1.0 to 2.0 MPa).
- From 6.00 ÷ 7.00 m depth from G.L. to about 10.00 m depth, medium to fine sand, with medium to high q_c values (from 7 to 15 MPa).
- Below 10.0 m depth, an alternation of silty clay, clayey silts and silty sands is found.

A typical CPTU profile around the bell tower is shown in Figure 8.

The maximum water level to be considered for the design of temporary works ("*chambers*") corresponds to the San Marco Square elevation around the area of the bell tower, that is approximately +0.90 m on msl.

This value was taken into account for uplift verification of the chamber for temporary conditions (during construction) as well as for final working conditions. Higher levels were not taken into account for uplift verification, being the chamber cover substantially open, allowing the flooding of chambers in higher water conditions ("*acqua alta*").

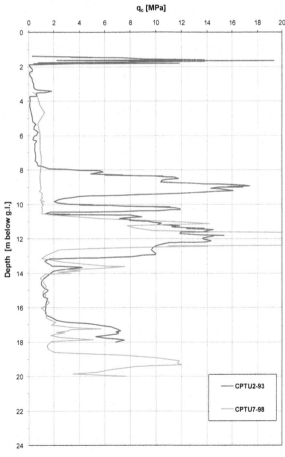

Figure 8.   Typical CPTU profile in the area of the San Marco bell tower.

235

4 THE PROJECT OF THE WORKS UNDER THE LEVEL OF THE PIAZZA

Deep cement mixing columns (DCM), reinforced with steel pipes were selected to provide water tightness, lateral support as well as bottom stability to uplift of the seven pits (*"chambers"*) that in turn will serve for the installation of the titanium bars. The chambers will be mostly below the water table, and shall remain accessible after completion of the works, to allow monitor of titanium bars and allow future interventions, if necessary.

Titanium rods will be installed in drilled holes through the foundation stone blocs and anchored at the foundation's corners by titanium plates, being initially prestressed to 250 kN each.

Depending on the foundation's structural responses, stresses in the titanium rods could be adjusted over a few years. Figure 9 shows a general 3D view of the project of the underground works, where the chambers are highlighted in orange, with a schematic view of deep mixing columns.

Corner chambers (Figure 10) have 10 to 15 m² footprint area, with different shapes according to surface constrains. Lateral walls of the chambers are formed with two or three rows of secant DCM (Deep Cement Mixing) columns, having diameter of 0.40 or 0.50 m, 8 to 9 m deep.

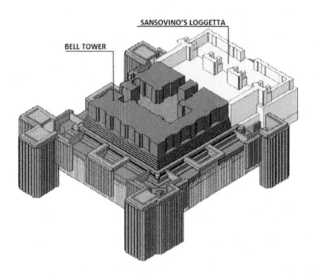

Figure 9.   3D model overview.

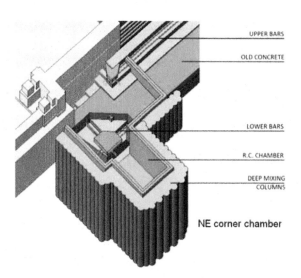

Figure 10.   Detail of 3D model of North East corner chamber.

From a geotechnical point of view, the main project issues were:

- Waterthightness of the chambers during excavation, to avoid any loss of fines due to water inflow;
- Stability of lateral walls of the chambers during excavation;
- Stability to uplift of chambers in temporary conditions taking into account the "high water";
- Control and limitation of total and differential settlements of existing foundations (bell tower and "Procuratie Nuove") to very small values, comparable to recorded seasonal movement due to tide, temperature, wind etc.;
- Control and limitation of vibrations induced by construction.

DCM technique was selected at an early stage of the project after a detailed review of pro and cons. In particular, the main advantages of the technique are the reduced vibration impact as well as the low disturbance to the soil surrounding the treated column during its formation.

Stability to uplift of corner chambers after completion is guaranteed by the dead weight of the r.c. chamber walls as well as of chamber cover. Stability in temporary conditions, after chamber excavation and during r.c. walls constructions, shall be guaranteed by the dead weight of the treated soil (taken as a massive block), as well as possible contribution of side friction against the alignment of deep mixing columns.

Strength of treated soil, in terms of UCS, was considered less critical than the density issue, being the expected stresses in treated soil columns during chambers excavation relatively low (<1.0 MPa).

5 EXECUTION OF THE WORKS

Both the construction of the underwater chambers, and the implementation of the 8 holes inside the foundation block required special equipments and skilled workers, given the unusual position and environment. Structural engineers worked to perforate the holes with such a precision that each bar should be introduced and later tensioned to its two opposite anchors without any contact with the foundation block. The holes for the lower chain were about 20 m long, and the needed precision was of few millimeters in order to introduce exactly the bars on the axis of the prestressing anchorage.

The system used was that of drilling beforehand a small hole with the help of laser guidance; afterward, the hole was gradually enlarged by means of a purposely built reamer, up to reach the diameter of 450 mm. The industrially produced bars have the maximum length of 3000 mm, so that the individual elements had to be connected by threaded cylinders. Strain gauges applied on the bars themselves have the function of load cells. They will be used for the initial tensioning of the bars at 250 KN, are working in full time for the monitoring of the tension and will be used eventually for keeping constant the thickness of the cracks of the foundation.

Such monitoring will be kept under control at least for 10 years and in every case up to a complete stabilization (Figure 11).

The introduction of the bars and of the anchorages, as well as their monitoring and possible modification, require the access of technicians to the anchorages of the bars, the control of the conservation of the geometry at the 4 corners of the foundation, materials, and eventually replacements or small modifications of the state of stress. This access is mandatory for the best control of the intervention and its conservation in the future. For this reason a not easy work has completed the intervention: the excavation and implementation of 4 underwater chambers at the 4 corners of the foundation.

Works under water, holes in concrete, consolidation of the soil and excavation of the chambers at the corners, reinforced concrete watertight walls and plates, positioning and tensioning of the bars after placing granite anchors, were only part of a delicate work aiming at avoiding also to damage the close buildings of Marciana Library, Procuratie, and residual archaeology of a Middle Ages previous building.

A small/medium size drilling rig (rather unusual for DCM columns) able to perform small diameter DCM was selected to minimize the site disturbance. Transportation of the rig (SM 21 SOILMEC) from the San Marco Basin to the bell tower required a special platform to spread out the weight of the

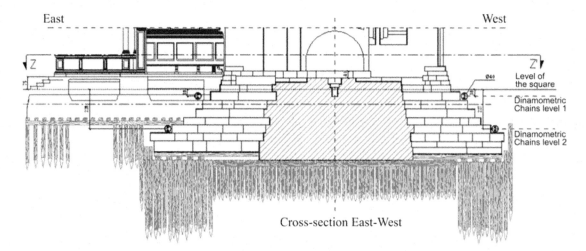

Figure 11. Dynamometric monitoring of titanium bars. E-W cross section.

rig (nearly 220 kN) and fulfill the Municipal regulation in force, that limits the maximum allowable pressure on the square paving "masegni" (trachyte grey stones) to 3 kPa.

The platform was built using a 150 mm (6″) layer of sand and a layer of neoprene, covered with wooden beams, over which steel I-beams were placed, blocked laterally with retainers anchored to the wooden beams. The 160 m route was covered in four stages of 40 m each. After each stage during the night the platform was disassembled, then reassembled in front of the machine. The entire stretch took less than a week to arrive at its destination (Figure 12).

The other major concern for the underground works was the effect on the existing monuments, in particular the bell tower and the nearby building of Procuratie Nuove with the Marciana Library, in terms of induced settlements and rotations.

The two concerned monuments have different foundations systems (Figures 13 and 14):

- The bell tower is founded on driven wooden piles, penetrating the sand layers below 6.0–7.0 m depth; the original foundation was reinforced with many additional piles during the reconstruction made in 1912.
- The Procuratie Nuove is on shallow foundations, with an expected depth that is slightly smaller than the design excavation level of corner chamber located on the South side of the bell tower. Along this side the overall strength of the intervention was improved by adopting 0.50 m diameter deep mixing columns and increased steel pipe reinforcement.

Preliminary laboratory and on site tests were carried out to assess the best cement content to obtain the strength required by the designer, and to assess the achievable density of the mixed soil.

Test results confirmed the possibility to achieve the required minimum compressive strength of 0.35 MPa at 28 days curing (on treated clay) and the minimum unit weight of 18 kN/m^3 (Figure 15).

Noise and vibrational full-scale tests to measure the vibrations generated by the rig during the installation of the columns were also performed, both at an outside test location (with similar soil profile) and, after positive results, at the project site before any site work. A dedicated vibration monitoring system was installed at the project site, to control the effective amplitude of generated vibration, as well as noise, during the whole duration of the works.

The preliminary noise and vibration tests on site confirmed that expected vibration level during deep mixing column construction was negligible and would not cause appreciable disturbance to residents and tourists.

The main works on the Piazza San Marco started in early August 2009.

Due to the possible presence of underground obstacles, the columns were drilled to the full depth using the minimum quantity of water; the cement slurry was added during retrieval at fixed speed.

Figure 12.   Rig crossing Piazza San Marco on the protection platform.

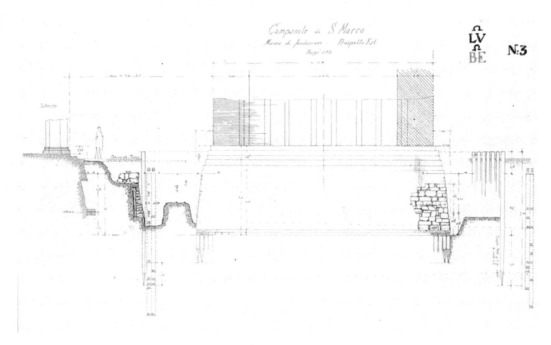

Figure 13.   Cross section showing the different foundation systems of the bell tower and of the Procuratie Nuove (from original sketches taken during the reconstruction).

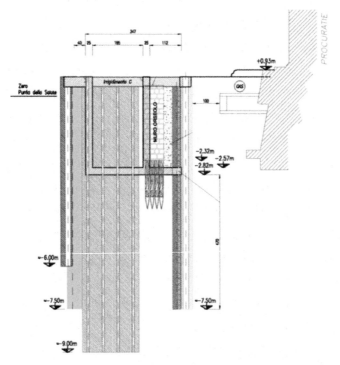

Figure 14. Cross section of SW chamber showing the foundations of Procuratie Nuove and of the buried Orseolo stone wall.

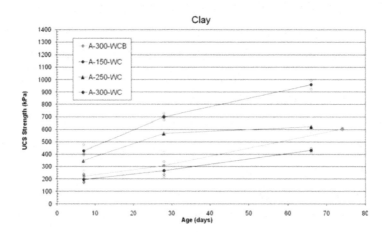

Figure 15. Results of early laboratory test on cay-cement mix.

As the lower half of the columns is in fine sand and the upper half in cohesive, soft, organic material, the tool was rapidly moved twice up and down to enhance the homogenization of the cemented fresh soil column. Special mixing tools have been developed to cope with the obstacles. During works, an adjustment of bottom plug thickness was required for some of the chambers, due to the unexpectedly low unit weight of treated soil detected from routine control tests, the reason of this being the big amount of wood chips (from buried wood pieces abandoned after historical works, Figures 16 and 17).

The columns bordering the chamber walls are reinforced with a 88.9 mm diameter steel pipe installed through the column's axis, after curing and re-drilling.

The spoil generated during the execution of the columns was collected and pumped to a filter press dewatering plant, which reduced its volume. The dried spoil was then transported by small cart to the quay, and boarded on pontoons to the final disposal area.

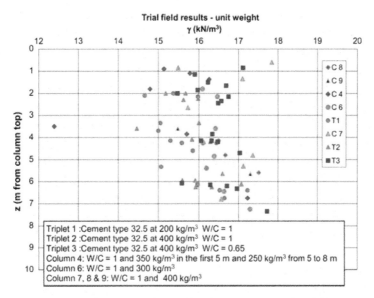

Figure 16. Trial field—unit weight of stabilized soil vs. depth for different tested mixes.

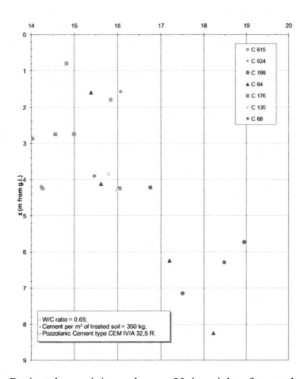

Figure 17. Project deep mixing columns. Unit weight of treated soil vs. depth.

The work progress was slowed down through periodic flooding. In fact the pavement of Piazza San Marco is at around +0.9 m above mean sea level, and, according to the agreed site procedures, works were halted whenever tides rose to +1.0 m elevation, for "acqua alta" (high tide, see Figure 18).

The excavation of the chambers started In August, 2010 and showed the general good quality of the job done, with very limited water inflow and very small movements of the monuments (bell tower and "Procuratie Nuove"), as shown by the monitoring system.

The monitoring system data and, in particular, the 30 m long-base pendulum installed inside the bell tower, shows that the movements of the bell tower during the intervention changed only few millimeters with respect to the usual seasonal variations.

Figure 18. "Acqua alta" (high tide) conditions during works.

6 CONCLUSIONS

The interventions on historical buildings and monuments always imply particular emphasis on the exploration of the subsoil to identify as much as possible the pre-existences in terms of buried structures, masonry, man-made fill, culverts, etc.

Generally this goal cannot be completely achieved before the site works, but a refinement of the investigation is needed with a multi-disciplinary approach, involving the archeological and historical expertise, in addition to the state-of-the-art geotechnical, geological and structural knowledge.

However uncertainties cannot be completely eliminated; therefore a comprehensive back-up and recovery plan shall be implemented at the beginning of the works.

The innovative strengthening intervention with the titanium pre-tensioned rebars requires very special skills and technologies, from the early investigation to the conceptual design and through all the consolidation works development, up to completion and maintenance.

The selected technology for the temporary works (deep in situ mixing columns) has proved its soundness in this very sensitive and valuable architectural context, allowing the completion of chambers with negligible displacements on the bell tower and the nearby monuments.

Working with titanium is much more demanding and challenging than working with stainless steel. It takes several months to order the material, to prepare the bars, the connections and the tensioning plates.

All these operations were achieved and controlled by December 2012 and now all the chains components are ready to be assembled. All the preparatory operations like archeological surveys, deep mixing, chambers building and drillings in the old concrete have been finished thus the intervention is estimated to be completed by January 2013.

Monitoring of the chains tension and of the response of the tower bell have been planned for the next three years.

REFERENCES

Municipality of Venice, 1912. Il Campanile di San Marco riedificato. Studi, ricerche, relazioni. *Editor C. Ferrari, Venice.*
Donghi, D. 1913. La ricostruzione del Campanile di San Marco a Venezia. *Giornale del Genio Civile, 1913.*
Macchi, G. 1993. Monitoring Medieval *Structures in Pavia. Structural Engineering International, 1/93 Int. Ass. For Bridge and Structural Engineers.*
Consorzio Venezia Nuova, 2006. Strengthening interventions on the foundation of San Marco Bell Tower. Final Design. *Venice.*
Vanni, D. Macchi, S. & Pastore, V. 2012. Deep Soil Mixing to help the restoration of Venice San Marco Bell Tower. *Deep Foundation Institute. 4th Intl. Conf. on Grouting and Deep Mixing. New Orleans, LA, USA.*

Geotechnics and Heritage – Bilotta, Flora, Lirer & Viggiani (eds)
© 2013 Taylor & Francis Group, London, ISBN 978-1-138-00054-4

Portus Julius: A complex of Roman infrastructures of late republican age

C. Viggiani
Emeritus Professor, University of Napoli Federico II, Napoli, Italy

ABSTRACT: The Greek Campanian architect L. Cocceius Aucto, active in the late Roman republican age, is believed to be the principal artificer of the complex of infrastructures serving the harbour of the Roman fleet, the *Portus Julius,* in the Phlegrean Fields east of Naples. They include three outstanding road tunnels that have been excavated in few years, starting on 39 B.C. under the impulse of Octavianus, who will soon become the first emperor Augustus, and his deputy and son in law Marcus Vipsanius Agrippa. The road tunnels have been practically invented by Augustus and Agrippa, but it was Cocceius to put their program in tangible form. In fact, the three Phlegrean tunnels are the longest and most impressive ones of the whole Roman civilization and represent unprecedented engineering masterpieces. The paper reports their main features, the geotechnical problems affecting them and the planned or carried out remedial interventions.

1   THE PHLEGREAN FIELDS

The Campanian sites East of Naples including Puteolis, Cumae, Baiae and the Lake Avernus, in the area of the so called Phlegrean Fields, are universally known and admired for the beauty of the landscape bearing the sign of the volcanic activity (figures 1 and 2), but also for the strong suggestion of legendary events (the entrance to the underworld through the Lake Avernus) and mythical figures (the hero Aeneas, the poet and wizard Virgil) and for the abundance of monuments and ruins (Hamilton, 1776).

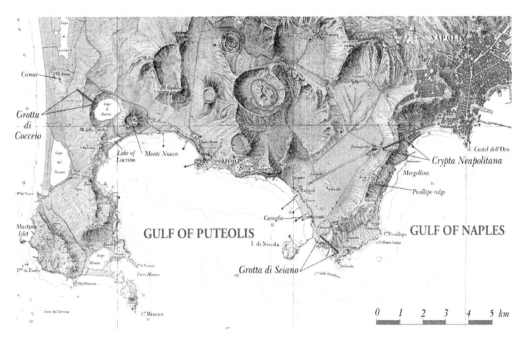

Figure 1.   The Phlegrean Fields in a map of the early XIX century.

The area has been inhabited since very ancient times. Cumae had been founded in the 8th century BC by Euboean Greeks, coming probably from the earlier colony of Pithecusae, on the nearby island of Ischia. The Greeks soon made Cumae the area's most powerful city, then spread down the coast, founding Parthenopes in ca. 680 BC, Dikaiarchia (Puteolis) in ca. 530 BC and Neapolis, the "New Town", in 470 BC. Cumae came under Roman rule in the 3rd century BC; its position was strategically important to control the coast from the mouth of the river Volturno to Puteolis. It is perhaps most famous as the seat of the Cumaean Sibyl; this prophetess was renowned for entering a trance, during which she was able to predict the future. In Virgil's poem Aeneid she predicted to Aeneas the glorious future of Rome.

In Roman times, and specially towards the end of the Roman Republic in the 1st century BC, Baiae was a fashionable resort for rich patricians, notorious for the hedonistic temptations and for rumors of scandal and corruption. There were sumptuous villas, with large swimming pools and domed casinos, including those of Julius Caesar and Nero; it was at his villa near Baiae that the emperor Hadrian died in AD 138. The Cento Camerelle (literally, hundred small rooms) are complex waterworks belonging to Ortensius Ortalo's villa.

Puteolis was established as a Roman colony in 194 BC; it became the most important commercial harbor of the Mediterranean and the great emporium for the Alexandrian grain ships. The volcanic pyroclastic sand is called pozzolana from Puteolis (Pozzuoli in modern Italian); it was the basic component of the famous Roman concrete. The columns of the Macellum, known as Temple of Serapis (fig. 3), show the holes of date shells (the marine mollusc Lithophaga Lithophaga), indicating the amount of past subsidence and upheaval. The Flavian Amphitheater is the third largest Italian amphitheater after

Figure 2.   A view of the Phlegrean Fields. On the left, Cape Misenum; in the background, the island of Ischia.

Figure 3.   The Puteolis Macellum, known as Temple of Serapis. The holes of date shells on the shaft of the three columns on the right reveals the level reached by the sea during cycles of subsidence and upheaval.

the Colosseum and the Capuan one, located in Campania also. In the middle of the Rione Terra, the acropolis of the ancient Puteolis, the remains of the temple of Augustus are incorporated into a church of the XVII century (fig. 4).

The Piscina Mirabilis (fig. 5) was a freshwater cistern dug entirely out of the tuff to provide the Roman fleet with drinking water. It is 15 m high, 72 m long and 25 m wide; its vaults are supported by 48 pillars; its capacity is 12,600 m^3; it was supplied with water by an aqueduct, the Aqua Augusta, that brought water from springs in Serino, 100 km apart.

The Lake Avernus (fig. 6) was considered by the Romans the entrance to the Hades (the underworld). In Virgil's account, Aeneas descends to the underworld through it.

Figure 4. The columns of the Temple of Augustus incorporated into a baroque church of the XVII century.

Figure 5. The Piscina Mirabilis.

Figure 6. The Lake Avernus.

245

2 PORTUS JULIUS

After the assassination of Julius Caesar in 44 BC, and the defeat of Cassius and Brutus at Philippi, the only opposition to Octavian (who will soon become the first emperor Augustus) was that of Sextus Pompey, son of Pompey the Great, who held Sicily with a powerful fleet. Octavian tried to conquer Sicily, but was defeated in the naval battle of Messina (37 BC) and again in August of 36 BC.

Marcus Vipsanius Agrippa, deputy and son in law of Octavian, reorganized the army and the fleet by installing an outstanding settlement in the Phlegrean Fields. Cumae was chosen as the main stronghold, and the littoral south of Cumae between the Cape Miseno and Monte di Procida was the seat of a military school; the site still keeps the name of Miliscola (Militum Schola). The Roman fleet was installed in Portus Julius, just in front of the Lucrino Lake; the shipyard of the fleet was established in the Lake Avernus. The remains of the structures of Portus Julius are at present submerged by the sea (Gunter, 1913), due to the volcanic subsidence of the area (fig. 7).

The establishment of Portus Julius was accompanied by the implementation of an outstanding complex of infrastructures (fig. 8):

- a tunnel, known as Grotta di Cocceio, from the North-West shore of Lake Avernus to Cumae;
- a tunnel across the acropolis of Cumae, known as Crypta Romana;
- a navigable channel connecting Lake Avernus to the lake Lucrino and the latter to the sea. Lake Lucrino is at present much smaller than in Roman times, because of the presence of Monte Nuovo, generated in one night by an eruption in the XVI century;
- a tunnel, known as Grotta della Sibilla, connecting the South-West shore of Lake Avernus to Baiae;
- a tunnel connecting Puteolis to Naples, known as Crypta Neapolitana (fig. 1);
- a tunnel connecting Puteolis to the Pausilypon villa, known as Grotta di Seiano (fig. 1).

Incidentally, Agrippa utterly defeated Pompey's fleet at Naulochus Cape in 36 BC. Pompey fled to the east and never re-established a position of strength; he was eventually killed in 35 BC.

The plans of Octavian and Agrippa in the Phlegrean Fields were implemented by a genial architect/ engineer, Lucius Cocceius Auctus.

Figure 7. The structures of Portus Julius submerged by the sea because of volcanic subsidence of the coast.

246

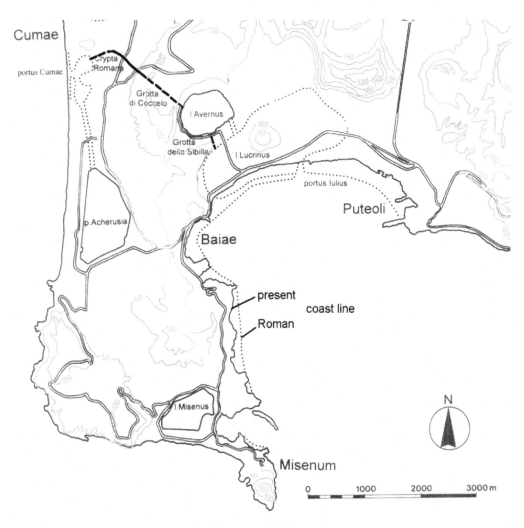

Figure 8.   The Phlegrean Fields at the Roman Age.

The role of an architect in the Roman civilisation (Gros, 1986) was completely different from the present one. He was not distinct from the engineer and, in any case, was not credited as the principal responsible of a work (a building, a monument, a bridge). In fact, the principal responsible was the *promotor,* the second the agent or contractor; the architect, if any, came only after these. They said: Caesar constructed that bridge. Only many centuries later they said: Perronet, Brunel, Calatrava constructed that bridge. This is the main reason why we know very little about the social status and the role of the architects at the Roman times.

The Greek—Campanian architect Lucius Cocceius Auctus is a notable exception. Cocceius was formerly the freedman of Lucius Cocceius Nerva, a rich and influent patrician belonging to the circle of Octavian. It was Nerva to obtain the first important assignments to his freedman. He introduced him into the narrow group of people involved in the great program of public works started by Octavian and Agrippa.

At the death of Nerva Cocceius, with some associates as Postumius Pollio, formed an outstanding group of designers and agents. They represent a fine example of social success, certainly connected to political support and to the availability of large capitals, but mostly to an outstanding practice of all the architectural and engineering specialities, making them capable of satisfying any kind of request.

As an architect, Cocceius constructed public buildings in Cumae, Formiae and the temple of Augustus on the acropolis of Puteolis, where his inscription has been found (fig. 9). As an engineer, Strabo and other classical historians quote him as the author of some of the major tunnels listed above: the

247

Figure 9. Cocceius' incription at the Temple of Augustus in Puteolis.

Grotta di Cocceio, the Crypta Neapolitana and, probably, the Grotta di Seiano also. Crypta is the Latin word and Grotta the medieval Italian word for any underground space, both natural or man- made.

3 THE TUNNELS

The main characteristics of the Roman tunnels in the Phlegrean Fields are listed in Table 1. All of them are excavated almost entirely through the Yellow Neapolitan Tuff, that is indeed an ideal material for tunneling being an easily excavated soft rock with a compressive strength ranging between 3 and 10 MPa and a friction angle of about 27°. The Phlegrean tunnels are by far the longest and most impressive tunnels of the whole Roman civilization; they are indeed unprecedented and unequalled engineering masterpieces.

The Grotta di Cocceio was rediscovered in 1844 (Scherillo, 1858) due to the collapse of the vault near the Cumae extremity, resulting in a sinkhole at the ground surface; it was re-excavated in the following years by will of the king of Naples Ferdinand II Bourbon and opened in 1861 by the king Vittorio Emanuele II, just after the unification of Italy. During the second World War it was used for the storage of explosives; in 1944 the retreating German Army blasted the explosives, producing at mid tunnel a huge cavity with a height of almost 40 m, that is now a further attraction for the visitors.

The Grotta di Seiano is an access to the famous Pausylipon, a villa belonging to the patrician Vedius Pollio and later to the emperor Augustus. It owes its name to Seianus, minister of the emperor Tiberius, who improved and enlarged it some fifty years after its construction. In the centuries, both the entrances were covered by landslides debris and the memory of the tunnel was lost; it has been rediscovered only in 1840, when Ferdinand II Bourbon became personally involved in its reopening, and committed the necessary works to Ambrogio Mendia, engineer and mathematician, Dean of the School of Engineering in Napoli (Lancellotti, 1840).

Reference to the Crypta Neapolitana may be found in Seneca (Epist. VI, 57, 1–2) and Petronius. In the Middle Ages, it was believed to have been excavated in only one night by Virgil, poet and wizard. The tunnel, in its present configuration, is the result of a number of later interventions; it has been in regular use as a road tunnel till 1885. At present, it is interrupted by collapses of the vault in the central stretch and some interventions are underway to re-establish the free transit.

The Crypta Romana connects the lower city of Cumae to the coast, passing under the acropolis of Cumae. The tunnel was believed to be a connection between the city and a hypothetical harbor, on the shore south of the acropolis. Recent archaeological investigations have proved that the Cumae harbor was actually located in the Lake of Licola, north of the acropolis. At present, hence, it is believed

Table 1. Characteristics of the tunnels.

Name	Length (m)	Width (m)	Height (m)	Ventilation/lighting shafts
Grotta di Cocceio	970	4.5	4.0–6.0	6 vertical, 2 inclined, 1 lateral
Grotta di Seiano	780	4.0–6.5	5.0–8.0	3 lateral
Crypta Neapolitana	711	4.5	4.6–5.2	2 inclined
Crypta Romana	293	5.0	6.0–8.0	6 vertical
Grotta della Sibilla	203	3.4–3.7	4.4	–

that the Crypta Romana was conceived to serve the quarters of soldiers and workers employed in the shipyard (Caputo, 2004).

The Grotta della Sibilla connects the Lake Averno to Baiae; near its western intake there is a lateral branch of the tunnel leading to some rooms which are at present submerged because of the subsidence. They are traditionally indicated as the seat of the ritual laver of the Sibyl.

The three major tunnels (Grotta di Cocceio, Grotta di Seiano and Crypta Neapolitana) have been affected in different periods by local collapses. Some engineering intervention of maintenance, upgrading and consolidation have been carried out, and more are needed to ensure safe fruition. They are dealt with in some details in the following paragraphs.

4 GROTTA DI COCCEIO

The Grotta di Cocceio (fig. 10), as reported above, connects the north-west shore of the Lake Avernus to the south-west sector of the ancient Cumae. It crosses the mount Grillo under a maximum cover of nearly 100 m and is mostly excavated in the yellow Neapolitan tuff. The border between the vault and the walls is marked by a horizontal step, revealing that the excavation was carried out in two stages: firstly in the vault, and afterward dawn to the floor. The final part towards Cumae, for a length of around 60 m, had been excavated through slightly indurated pozzolana and was originally provided by a concrete (*opus cementicium*) vault, that collapsed in the XIX century revealing the existence of the tunnel. At present, this stretch is an open trench with steep slopes in pozzolana and pumices.

The floor of the tunnel has an elevation of 1.3 m above sea level at the Avernus intake, and 42 m a.s.l. at Cumae; there is a slope of 5.4% for 800 m on the Avernus side, then an horizontal stretch about 70 m long, and finally a 100 m long stretch sloping toward Cumae. It was enlightened and ventilated by 9 shafts, six of which vertical, two inclined and one lateral. The last two vertical shafts on the Cumae side were excavated entirely in pozzolana, and have been cancelled by the collapse of the roof and the transformation of the tunnel in a trench.

The four remaining vertical shafts have either a square section of 4×4 m² in plan, or a rectangular section of 4×8 m² with the long side parallel to the tunnel axis. In the upper part, crossing pozzolana, the shafts are lined by masonry of tuff stones with irregular shape (*opus incertum*).

The two inclined shafts have a stepped floor, since they had also the purpose of access to intermediate excavation faces during the construction of the tunnel.

The lateral shaft, oriented westwards and enlightened in turn by a vertical shaft some meters after the branch, had probably also the function of way to the amphitheater, located south of the urban walls of Cumae.

In the I century AD the whole system of infrastructures of Portus Julius was converted to civil use, and so did the Grotta di Cocceio; it had been directly connected to the new Via Domitiana.

In the Middle Ages Cumae gradually became a bandits and pirates lair, and therefore it was finally destroyed in 1207 to guarantee the safety of Napoli. The tunnel was abandoned and gradually filled by the sediments entering from the shafts; the two intakes were covered by landslide debris and the memory of the monument was lost until its rediscovery in 1844.

As mentioned above, the central part of the tunnel was interested in 1943 by an explosion, producing a huge cavity with a height of 37 m, a width of 20 m, a length of 60 m and a volume of around 18,000 m^3 (fig. 11). After the war the monument was closed by walls at the two intakes; in spite of that, people searching for war surplus entered through the shafts. A consequence of this incautious commerce was another explosion in 1951, killing five young men and producing two further smaller cavities on the Cumae side.

After 1980 the tunnel has been the object of gradual interventions aimed at reopening it to public. The explosion debris and some unfired explosives have been removed, and in the 1990's the whole length was unobstructed and could be traveled over. The fruition of the great central explosion cavity, however, was unsafe due to the risk of rock blocs falling from the roof and walls and even a possible collapse of the roof. A detailed investigation of the stability conditions of the cavity has been carried out in 2000. It included a number of boreholes drilled from the ground surface to recover intact

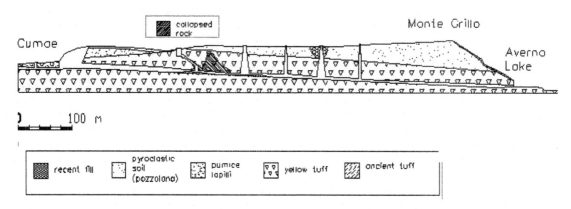

Figure 10.   Longitudinal profile of the Grotta di Cocceio.

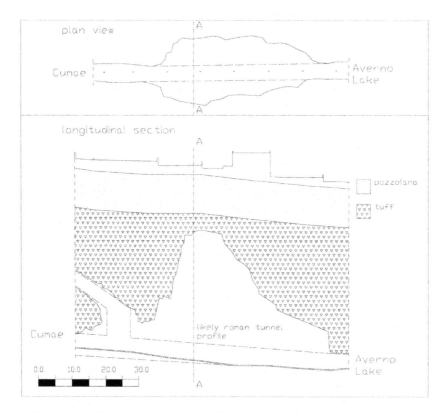

Figure 11.   Grotta di Cocceio. Plan view and longitudinal section of the huge cavity produced by the blasting, superposed to the original shape of the Roman tunnel.

sampled of the pyroclastic cover soils and cores of the tuff. The geotechnical characterization of the materials has been obtained by laboratory and site tests. A survey of the cavity by laser scanner completed the investigations.

A detailed mapping of the discontinuities on the sidewalls and roof could not be obtained for safety reasons, but some indication on joints and fractures were gained by a careful inspection of the photos and cores.

A family of subvertical discontinuities, roughly normal to the tunnel axis and apparently pre-existing the explosion, has been found. Furthermore, two of the boreholes intersected a horizontal discontinuity in the tuff above the cavity vault, 3 m beneath the top of the tuff, probably generated by the blasting in correspondence of a pre-existing weak layer, such as a level of pumices embedded in the tuff.

A FEM analysis of the cavity (fig. 12), trying to simulate at best all the listed features, has been carried out by Amato *et al.* (2001); they concluded that the cavity is globally stable, but the tuff located between the roof and the horizontal discontinuity is unstable and subject to fall. They claim that a progressive failure, leading in the long term to a sinkhole, could possibly occur following the fall of blocks.

In principle, the occurrence of a sinkhole is not a tragedy. The vault of the cavity could be left free to collapse, producing a further ventilation shaft and reproducing the amazing view of the sun beams entering the nearby *Piscina Mirabilis* (fig. 5). The visitors' safety could be obtained by just excavating a pedestrian tunnel around the explosion cavity, bypassing it but allowing its view from the intersections with the Roman tunnel. Unfortunately this appealing solution had to be abandoned because of the occurrence of some buildings above the cavity, even if these buildings had been constructed without any authorization and are hence totally irregular.

Different solutions have been considered to solve the problem, as that of filling the cavity with lean concrete and then re-excavating the tunnel through the fill, or that of protecting the visitors with a steel canopy. The problem is further complicated by the fact that a colony of chiroptera (bats) has elected the cavity as nest, and they are a protected species and hence they cannot be disturbed by the works. This problem, epitomizing the complexity of conservation, has been discussed at length and is not yet completely solved.

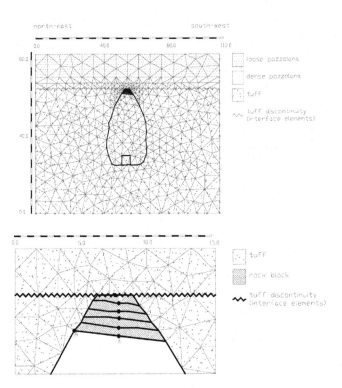

Figure 12. Grotta di Cocceio. (Above) Finite element model of the explosion cavity; (Below) Detailed mesh around the vault of the cavity.

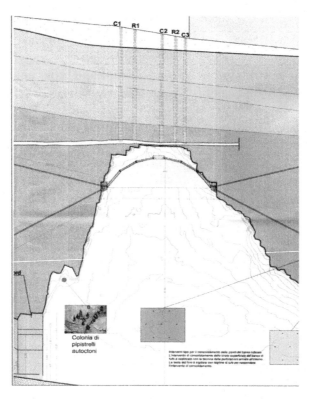

Figure 13.   Stabilization of the vault of the explosion cavern.

Figure 14.   Grotta di Cocceio; the results at surface of the collapse of a lighting shaft.

Eventually, it was decided to line the upper part of the roof with a steel ribs dome, fixed to the tuff by micropiles (fig. 13), filling the space between the dome and the rock with cement grout or polyurethane foam, in order to implement an effective prevention measure against the detachment and fall of blocks; the sidewalls of the cavity will be put in safety by detaching the threatening blocks, sealing the fissures and locally fixing larger blocks by nails.

In 2009, while starting those works, one of the vertical ventilation shafts some 100 m away from the explosion cavity on the Averno side suddenly collapsed for the failure of the masonry linings in the upper part; the effects can be seen in figs. 14. The tunnel was interrupted again, and a building at the surface was endangered by a possible spreading of the collapse. It was decided to renounce, at least temporarily, to re-establish the shaft; the depression at the surface has been filled, the debris within the shaft has been consolidated from the surface by jet grouting, and the tunnel will be re-excavated through the debris. After this event, a careful inspection of the remaining shafts has been carried out, and some preventive consolidation of the masonry is underway.

We owe to our German friends the amazing explosion cavity in the mid of the Grotta di Cocceio, and in the present economic crisis the European rigor policy and the consequent lack of resources for such esoteric and unproductive things as an ancient Roman tunnel. But we know that …. "people will not look forward to posterity, who never look backward to their ancestors" (Burke, 1790). It is hoped, therefore, that in a near future the tunnel will be opened to public, allowing the fascinating 15' walk from the Lake Averno shore to the ancient Cumae.

5 GROTTA DI SEIANO

The bay of Puteolis is separated from the bay of Naples by the Posillipo hill, a tuff ridge with steep slopes and the top at a fairly constant elevation of 150–160 m above sea level, extending from north-east to south-west till the sea (fig. 1). Near the south-west extremity of the ridge, on the Naples side, there was the famous *Pausilypon*, a villa belonging to Vedius Pollio and later to the emperor Augustus (fig. 15). The name derives from the ancient Greek and means "painless"; it is at the origin of the modern name, Posillipo, of the whole area.

The Grotta di Seiano crosses the Posillipo ridge to connect the villa to Puteolis. Some Authors (Gunter, 1913) suggested that it was part of a coastal road ending in Naples, but this hypothesis has not been confirmed by archaeological investigations; the tunnel was probably just an access to the villa. A profile of the tunnel is reported in fig. 16. It is provided with three lateral shafts, opening over the small bay of Trentaremi (fig. 18); their features are reported in Table 2.

Probably the main function of these lateral ducts was ventilation, since the third one, the longest, is not rectilinear and hence it brings practically no light to the main tunnel. On the other hand, the main tunnel is known to be infested by poisonous gas making the ventilation particularly important.

The approach of Romans to the design of tunnel linings is well described by the following quotation from Vitruvius: "And if tuff or rock will be found, then a simple excavation will be executed through

Figure 15.   Pausilypon, the villa of Vedio Pollio and later of the emperor Augustus.

them; if on the contrary a sandy cohesionless soil will be found, then walls and vault will be provided in order to benefit the excavation".

(*Et si tofus erit aut saxum, in suo sibi canalis excidatur; sin autem terrenum aut harenosum erit solum, et parietes cum camera in specu struantur et ita perducantur*. Vitr., VIII, 6.3). In the case of the Grotta di Seiano, the eastern part of the tunnel is excavated through competent tuff and is hence unlined over a length of 144 m. The central and western parts cross slightly indurated pozzolana and very weak tuff; the sidewall of the tunnel in this section are lined with masonry in *opus incertum and opus reticolatum*, while the vault is lined with concrete (*opus cementicium*). The thickness of the masonry is around 0.5 m while that of the vault ranges between 0.5 and 0.8 m.

When the tunnel was rediscovered in 1840 it could be entered only by one of the lateral ducts; a number of collapses of the roof were found in the central stretch. Ambrogio Mendia reports the occurrence of ten major collapses, that were successfully restored after removing the debris, in spite of the difficulties connected to the occurrence of poisonous gases (Lancellotti, 1840). The masonry lining was reinforced and thickened over a number of stretches, totaling around 250 m, and a total of 68 masonry arches were added. In the reinforced sections the width of the tunnel is reduced to 2.6 m (fig. 17, 18).

At present the tunnel is open and regularly used as the principal access to the archaeological park of Pausylipon.

Table 2.   Lateral shafts to the Grotta di Seiano.

Distance from the east intake (m)	Length (m)	Width (m)	Height (m)
66	40	1.6	2.7
183	28	1.6	2.6
294	197	1.2	2.1

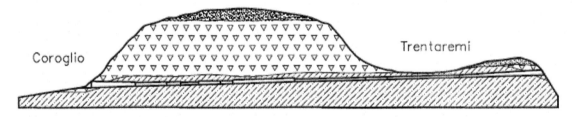

Figure 16.   Longitudinal profile of the Grotta di Seiano.

Figure 17.   Grotta di Seiano. (Left) The outlet of one of the ventilation tunnels; (Right) The entrance of the main tunnel on the Pozzuoli side.

Figure 18.   The arches added in XIX century on the western part of the tunnel.

6   CRYPTA NEAPOLITANA

As mentioned before, in the first century BC Puteolis had reached the apex of its military and economic importance, being the largest commercial harbour of the western Mediterranean sea. The city was connected to Neapolis and to the rich villas of the Neapolitan coast between Mergellina and Megaris (the present Castel dell'Ovo) by an old way through the hills (*via per colles*) dating back to the settlement of the first Roman colony in Puteolis, in 194 BC.

The Crypta Neapolitana was excavated across the Posillipo ridge to make easier and faster the communications between the two cities. The longitudinal profile of the tunnel is reported in fig. 19; it is 711 m long and oriented in east-west direction. There are two inclined ventilation shafts; the eastern one originates in the tuff cliff just above the eastern entrance and is 145 m long and inclined 6.2% over horizontal. The western shaft is 100 m long and inclined 44% over horizontal; it originates at mid slope of the hill and reaches the vault of the main tunnel at a distance of 100 m from the western intake.

The original Roman cross section was 4.5 m wide, to allow the passage of two carts, and probably 3.5 to 4 m high, with cylindrical vault and vertical side walls. The present configuration is the result of a number of later interventions. The first documented one was aimed at widening the tunnel and lowering the eastern intake to make easier the access; it was promoted in 1445 by the king Alphonse of Aragòn. A century later the viceroy Pedro de Toledo carried out further lowering of the road level in the eastern part and paved the roadway.

Further works were carried out in 1748 by the king Charles III Bourbon (Gunter, 1913) and in 1893 by the Municipality of Naples. During these latter works the eastern third of the tunnel was strengthened by about 100 pointed masonry arches.

Following a continuous deterioration of the statical conditions, the tunnel was closed to transit in 1885.

In 1930 the area surrounding the eastern entrance was re-shaped to accommodate a public park, including a Roman columbarium believed to be the tomb of Virgil, and the tomb of the great Italian poet Giacomo Leopardi. At that time the eastern stretch of the tunnel was partly filled, to rise the floor by as much as 9 m. In the meantime the side walls were braced by a number of masonry aprons, immersed in the fill (fig. 20, Chierici, 1929). Amato *et al.* (2001) have reconstructed the history of successive lowering of the tunnel floor using as a marker the grooves left in the tuff walls by the hubs of the carts wheels; such a reconstruction is reported in fig. 21 for a section 20 m distant from the eastern intake.

The elevation of the floor at the time of the maximum lowering was 25 m above sea level; the Aragonese floor was 37.9 m a.s.l. and the Roman floor 40.9 m a.s.l. The maximum lowering has been about 16 m. On the longitudinal profile the floor of the Roman tunnel was gently rising westward with a slope of 0.85%; after the successive lowering steps, the slope increased to over 3%.

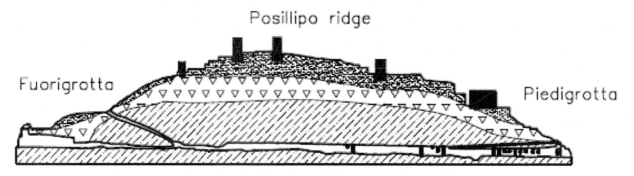

Figure 19. Crypta Neapolitana; longitudinal profile.

Figure 20. Masonry apron at the eastern intake of the Crypta, to be submerged later in the fill provided in 1936 to raise the elevation of the tunnel floor.

256

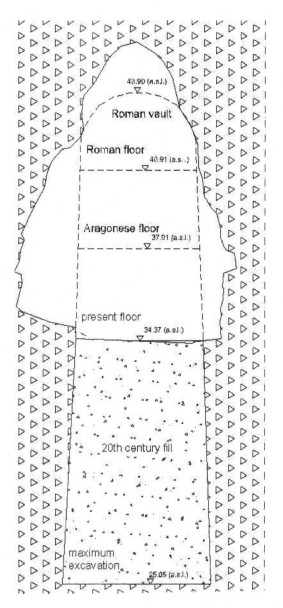

Figure 21. Crypta Neapolitana; reconstruction of the elevation of the floor in the section at 20 m from the eastern intake.

At present the tunnel can be travelled over a length of about 140 m from the eastern entrance and 200 m from the western one; the central section, over a length of about 370 m, is interested by substantial collapses and cannot be travelled.

As a consequence of collapses and excavation the present width ranges from 5 m to 12 m and the height from 5 m to 14 m. In the eastern accessible section there are the remains of 9 pointed masonry arches of the XIX century, all completely collapsed. To design remedial works, a broad investigation has been recently carried out, including a number of boreholes from the ground surface and core drills in the side walls of the tunnel and the eastern ventilation shaft. Measurements of the in situ stress by flat jack around the opening have also been carried out.

The values of the uniaxial compressive strength of the tuff along the tunnel axis are reported in fig. 22, together with the overburden pressure and the values of the vertical stress obtained by flat jack measurements and by a number of FEM analyses. The analyses show that the tunnel is globally stable, and widening and lowering the floor exert an almost negligible influence. A significant volume of tuff around the opening, however, has been found at yield (fig. 23); if it is assumed that the yielded volume

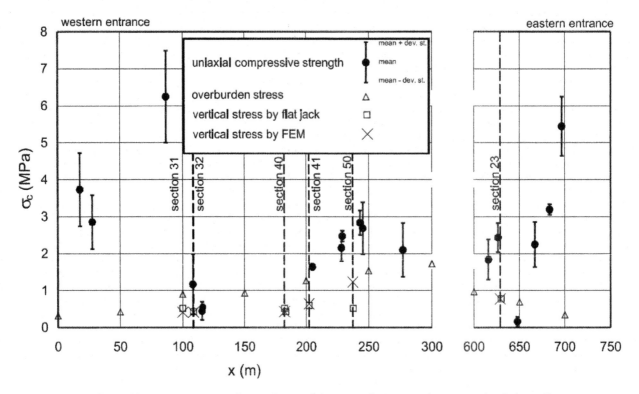

Figure 22. Crypta Neapolitana: State of Stress and compressive strength of the tuff.

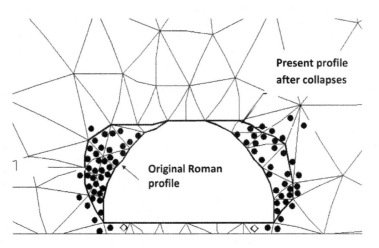

Figure 23. Comparison between the original Roman cross section, the one resulting after partial collapse and the results of FEM analysis.

progressively collapsed, the tunnel section reshapes in good agreement with the present tunnel section resulting from blocks breakdown and collapse.

7 CONCLUDING REMARKS

The complex of infrastructure realized in a few years in connection with the project of Portus Julius are a monument to the planning vision and the engineering capabilities of the Romans and to the architect (the engineer?) Lucius Cocceius Aucto.

At present, only the Grotta di Seiano is totally accessible. It is included in the archaeological park of Posillipo and crossed to enter the remains of the Pausylipon, the ancient villa of Vedius Pollio and later

258

of the emperor Augustus. The distress in the lighting shaft of the Grotta di Cocceio and in the large explosion cavity are being repaired, and hopefully in the near future a walkway from the Avernus lake to the acropolis of Cumae will be opened. The feasibility of reopening to public the Crypta Neapolitana also is being explored. The preliminary analyses carried out seem to indicate that the tunnel can be stabilized without massive and intrusive structures.

Modern Geotechnical Engineering is playing a major role in the conservation of these masterpieces of ancient Geotechnical Engineering.

ACKNOWLEDGEMENTS

Lucio Amato, Paolo Caputo and Clemente Esposito have generously provided data and suggestions respectively for Crypta Neapolitana, Grotta di Cocceio and Grotta di Seiano; their contribution is gratefully acknowledged.

REFERENCES

Amato, L. Evangelista, A. Nicotera, M.V. & Viggiani. C. (2001). *The tunnels of Cocceius in Napoli: an example of Roman engineering of the early imperial age.* AITES-ITA 2001 World Tunnel Congress, Milano, vol.I, 15–26.

Burke, E. (1790) *Reflection on the revolution in France.* Dodsley, London.

Caputo, P. (2004) *La Grotta di Cocceio a Cuma: nuovi dati da ricerche e saggi di scavo.* Viabilità e insediamenti nell'Italia Antica. L'Erma di Bretschneider, 309–330.

Chierici, G. (1929) *Il consolidamento della tomba di Virgilio.* Bollettino d'Arte Min. P.I., Bestetti a Tumminelli, Roma IX, 1, 438–455.

Gros, P. (1986) *Status sociale et role culturel des architects (période hellenistique et augustéen).* Architecture et Societé de l'archaism grec à la fin de la république romaine. Actes du Colloque International, Rome, 425–452.

Gunter, R.T. (1913) *Pausilypon, the imperial villa near Naples.* Oxford University Press.

Hamilton, W. (1776) *Campi Phlegraei.* Fabris, Napoli.

Lancellotti, L. (1840) *Sullo scavo della Grotta di Seiano e sulla nuova strada di Coroglio: cenno artistico letterario.* Napoli.

Scherillo, G. (1858) *Della meravigliosa spelonca Romana tra l'antica città di Cuma e il lago d'Averno.* Bollettino Archeologico Napoletano, n.s. VI, 173.

Geotechnics and Heritage – Bilotta, Flora, Lirer & Viggiani (eds)
© *2013 Taylor & Francis Group, London, ISBN 978-1-138-00054-4*

Author index